Essential Maths Skills

for AS/A-level

Design and Technology

Peter Warne and Chris Walker

HODDER
EDUCATION
AN HACHETTE UK COMPANY

Hodder Education, an Hachette UK company, Blenheim Court, George Street, Banbury, Oxfordshire OX16 5BH

Orders

Bookpoint Ltd, 130 Park Drive, Milton Park, Abingdon, Oxfordshire OX14 4SE

tel: 01235 827827

fax: 01235 400401

e-mail: education@bookpoint.co.uk

Lines are open 9.00 a.m.–5.00 p.m., Monday to Saturday, with a 24-hour message answering service. You can also order through the Hodder Education website: www.hoddereducation.co.uk

ISBN 978-1-5104-1706-9

First printed 2018
Impression number   5
Year   2022 2021 2020

Typeset in India
Cover illustration: Barking Dog Art
Printed in India

Hachette UK's policy is to use papers that are natural, renewable and recyclable products and made from wood grown in sustainable forests. The logging and manufacturing processes are expected to conform to the environmental regulations of the country of origin.

Contents

Introduction

Mathematical skills are vital in design and technology. Not only will they make up at least 15% of the marks in your exams but they are also essential if you wish to continue your studies beyond A-level into design- or engineering-related careers. This book covers all the maths skills you are required to know for your exams and it will be useful throughout the NEA. The book assumes little prior knowledge and all concepts are explained thoroughly, so if you aren't feeling confident with concepts that you were taught when you were younger, there is no need to worry. In each section the sets of questions start easy and build in difficulty.

All the questions in the book relate to design and technology, so not only are you practising maths skills, you are also developing your subject knowledge. Most questions will involve the use of multiple maths skills. Practice with these questions will help you to identify the links between skills and new contexts that you will encounter in the exams. There are also exam-style questions at the end of each section that integrate different maths skills and design and technology subject knowledge.

The content for the Design Engineering only sections is available online at **www.hoddereducation.co.uk/essentialmathsanswers**

1 Using numbers and percentages

In this chapter you will learn how to work with numbers and apply mathematical techniques that you will experience throughout your design and technology course. Many of the fundamentals of working with numbers are covered and you should ensure that you know these skills and practise them before moving on to later chapters.

1.1 Units, powers, standard form and accuracy

In design and technology, most quantities that you work with have an associated unit. In most cases an exam question will identify the units that you should use, but if it does not, you should ensure that you include the unit for your answers to be complete and correct. It is not necessary to show the units in each line of the workings of your calculations, but they should be given with the final answer.

SI units

SI stands for *Système international*, which is the **International System of Units**. The SI is a globally agreed system of units of measurement, with seven base units. Where possible you should use them and convert quantities to these base SI units. We will look at conversion later in this section. The SI units that you will encounter with greatest frequency are shown in Table 1.1.

Table 1.1 The most frequently encountered SI base units

Base quantity	Name of unit	Symbol
length	metre	m
mass	kilogram	kg
electrical current	ampere	A
time	second	s

The metre, ampere and second are usually seen with prefixes that represent the quantity in powers of 10. For example, 1 centimetre is 1.0×10^{-2} metre. The prefix 'centi' tells us that there are 100 of these units in 1 metre.

Millimetres are the most common measurement of length used in design and technology, engineering and architecture. The prefix 'milli' tells us that there are 1000 of these units in 1 metre.

The base quantity of mass, the kilogram, is the odd one out as it already has a prefix associated with it in the base unit. 'Kilo' means 1000 or 1.0×10^3.

You will learn more about prefixes later in this chapter. Before that, however, it is worth looking at standard form, which can help to identify the correct prefix. Standard form uses powers, which are explained in the next section.

Powers and roots

In design and technology, you will frequently come across **powers**, such as squaring and cubing, and **roots**, such as the square root and the cube root. For example, when the area of a circle is calculated, you will square the radius, or when the side of a triangle is being calculated, you may use Pythagoras' theorem to calculate the square root of the sum, or difference, of the square of the other sides. When calculating volume, the resulting units are likely to be m^3. In Chapters 3 and 4 you can work through these specific skills. However, before you do, it is essential that you understand what these terms mean and are confident in using them.

Powers

For the number written as 2^3, the small number is known as the power. The power number tells you how many times to multiply the number. In this case the power is 3 and so 2 is multiplied 3 times. Table 1.2 shows examples of powers.

Table 1.2 Examples of powers

Number	Means...	Which is...
2^2	'two to the power of two'	$2 \times 2 = 4$
2^3	'two to the power of three'	$2 \times 2 \times 2 = 8$
2^4	'two to the power of four'	$2 \times 2 \times 2 \times 2 = 16$

For any number x, with a power of n, the following applies:

x^n means 'x to the power of n' which is x multiplied n times.

A **square** number is any number that is the result of multiplying a number by itself. For example, 16 is a square number because $4 \times 4 = 16$. 16 can therefore be rewritten as 4^2. This is normally called 'four squared'. For any number, x, that is multiplied by itself, x^2, it can be called 'x squared'.

On a scientific calculator or scientific calculator app, this is normally the 'x^2' button. You should press this button after entering the value of x. For example, for 30^2, the order should be: type 30, then press 'x^2'.

Cube numbers have been multiplied three times. For example, 27 is a cube number because $3 \times 3 \times 3 = 27$, which can therefore be rewritten as 3^3. This is normally called 'three cubed'. For any number, x, that is multiplied three times, x^3, it can be called 'x cubed'.

Most popular scientific calculators and apps show this as the 'x^3' button. Again, you should press this button after entering the value of x.

For **powers greater than three** it is normal to say that the number has been multiplied **'to the power of'**. On your calculator, this is a little trickier. Many scientific calculators and apps have 'x^y' or a button as shown in Figure 1.1 with the power denoted as a square.

The button presses are as follows: type the value of x, then press the button 'x^y' followed by the value of y. For example, for 4^5, type in 4, then 'x^y', followed by 5.

Roots

The most common root that you will need to calculate in design and technology will be the square root. This generally occurs when using Pythagoras' theorem to calculate dimensions of designs, components or materials that involve the use of right-angled triangles. You can explore this in more detail in Chapter 4.

Cube roots are less common but you may come across them when calculating dimensions from the volume of an object or a material.

Square roots and cube roots are the opposite of squaring and cubing respectively.

- The symbol for square root is $\sqrt{}$.

- The symbol for cube root is $\sqrt[3]{}$.

- For any root, n, the symbol for the root is $\sqrt[n]{}$.

Examples

The square root of 64, written as $\sqrt{64}$, is 8 because $8^2 = 8 \times 8 = 64$.

The cube root of 8, written as $\sqrt[3]{8}$, is 2 because $2^3 = 2 \times 2 \times 2 = 8$.

$\sqrt[5]{32} = 2$ because $2 \times 2 \times 2 \times 2 \times 2 = 32$.

Scientific calculators and apps have buttons for square and cube roots and a special button for roots higher than three. These operate in a similar way to the power buttons, but the main difference is that it is usually necessary to activate some of them by accessing the calculator's second bank of functions, usually activated by pressing 'shift' or '2nd'. Figure 1.1 shows how to access these functions on a popular scientific calculator.

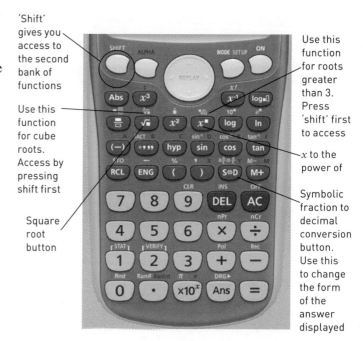

'Shift' gives you access to the second bank of functions

Use this function for cube roots. Access by pressing shift first

Square root button

Use this function for roots greater than 3. Press 'shift' first to access

x to the power of

Symbolic fraction to decimal conversion button. Use this to change the form of the answer displayed

Figure 1.1 Powers and roots on a scientific calculator

To calculate a square root, first press the square root button, then enter the value followed by '='. For a cube root, use the cube root button. For roots higher than three, the button presses are a little trickier. For example, for $\sqrt[5]{32}$, first press the root button, then enter the root value 5, followed by the number 32, then '='.

When working with decimals your scientific calculator may present the result as a fraction, for example $0.1^2 = \frac{1}{100}$. You may know that this is 0.01, but for more complex fractions you may need to use the calculator to convert them to a decimal answer. On many popular calculators, if you press the calculator button 'S $\leftrightarrow$ D', the answer will be converted from a symbolic fraction to a decimal.

(A) Worked examples

Calculate the following powers and roots.

a 24²

Calculator steps:

1   Type in 24.

2   Press 'x^2'.

3   Press '='.

Answer displayed: 576

b 0.2²

As a calculator will be used here, you can work with the value in the decimal form.

Calculator steps:

1   Type in 0.2.

2   Press 'x^2'.

3   Press '='.

Answer displayed: $\dfrac{1}{25}$

4   Press 'S ↔ D' button.

Answer displayed: 0.04

c 4³

Calculator steps:

1   Type in 4.

2   Press 'x^3'.

3   Press '='.

Answer displayed: 64

d 0.0003³

As a calculator will be used here, you can work with the value in the decimal form. However, if there were any more zeros before the first digit, it is best practice to convert the value to standard form to avoid errors (see next section).

Calculator steps:

1   Type in 0.0003.

2   Press 'x^3'.

3   Press '='.

Answer displayed: 2.7×10^{-11}

This is an answer in standard form.

Standard form and powers of 10

Standard form is a method of writing down easily very large or very small numbers. 1.2×10^6 kg is a value of mass in standard form. This represents the value 1 200 000 kg because $1.2 \times 10^6 = 1.2 \times 10 \times 10 \times 10 \times 10 \times 10 \times 10 = 1\,200\,000$. A number in standard form has one significant figure before the decimal point and a power of 10 following it. In this case 1 and 10^6 respectively. Table 1.3 shows the most common powers of 10 that you need to know.

There is an easy way to convert large numbers into standard form powers of 10: count the digits that follow the first significant number. This equals the power of 10. For example:

- 10 000 has four zeros after the '1', so the power of 10 is 4 and the number is 1.0×10^4 in standard form

- 3 450 000 has six digits after the 3, so the power of 10 is 6 and the number is 3.45×10^6 in standard form

Table 1.3 Positive powers of 10

Number	Standard form
10	1.0×10^1
100	1.0×10^2
1 000	1.0×10^3
10 000	1.0×10^4
100 000	1.0×10^5
1 000 000	1.0×10^6

Notice how, in the large numbers above 1000, the digits are separated into groups of three by a space. This is standard convention and avoids confusion that is often caused by using a comma instead of a space.

Getting into the habit of writing large numbers in this way makes conversion into standard form much easier because each group of three zeros, '000', is equal to 10^3.

Standard form uses negative powers of 10 to represent fractions, as shown in Table 1.4. The force 0.00034 N written in standard form is 3.4×10^{-4} N.

Table 1.4 Negative powers of 10

Fraction	Decimal	Standard form
$\frac{1}{10}$	0.1	1.0×10^{-1}
$\frac{1}{100}$	0.01	1.0×10^{-2}
$\frac{1}{1000}$	0.001	1.0×10^{-3}
$\frac{1}{10\,000}$	0.0001	1.0×10^{-4}
$\frac{1}{100\,000}$	0.00001	1.0×10^{-5}
$\frac{1}{1\,000\,000}$	0.000001	1.0×10^{-6}

There is an easy way to convert small numbers into standard form powers of 10: count the digits to the right of the decimal point up to and including the first significant digit. This equals the negative power of 10. For example:

- for 0.0001 the 1 is four digits to the right of the decimal point, so the power of 10 is −4 and the number is 1.0×10^{-4} in standard form

- for 0.00000345 the 3 is six digits to the right of the decimal point, so the power of 10 is −6 and the number is 3.45×10^{-6} in standard form.

Multiplying and dividing powers of 10

Multiplying and dividing powers in standard form is simple if you remember the following rules:

- To multiply powers of 10 you add the powers.

- To divide the powers of 10 you subtract the powers.

For example:

$$10^4 \times 10^2 = 10^6$$

In this multiplication, the powers 4 and 2 have been added to make 6.

$$\frac{10^4}{10^3} = 10^1 = 10$$

In this division, the power 3 is subtracted from 4 to give 1.

The following is also true:

$$\frac{1}{10^3} = 10^{-3}$$

Dividing by the power is the same as multiplying by the negative power, so for any power, n:

$$\frac{1}{10^n} = 10^{-n}$$

Remembering this enables you to treat divisions of powers as multiplications. For example:

$$\frac{10^4}{10^2} = 10^4 \times 10^{-2} = 10^2$$

 ## Worked examples

a **Write the following as powers of 10:**

   i **100 000**

   Count the zeros: five. This indicates the power of 10 should be 5.

   $100\,000 = 1.0 \times 10^5$

   ii **0.000005**

   Count the positions backwards from the decimal point and include the first significant digit (5). The count is −6, which is the value of the power.

   $0.000005 = 5 \times 10^{-6}$

b **Write the following in standard form:**

   i **0.36 mm**

   A number in standard form should have a single significant number ahead of the decimal point. The zero is not significant.

   $0.36\,\text{mm} = 3.6 \times 10^{-1}\,\text{mm}$

   ii **540 000 N**

   $540\,000\,\text{N} = 5.4 \times 10^5\,\text{N}$

   iii **0.00354 kg**

   The zeros are not significant.

   $0.00354\,\text{kg} = 3.54 \times 10^{-3}\,\text{kg}$, which is also known as 3.54 g

c **Multiply the following numbers without a calculator:**

   i **10 000 and 100 000**

   Convert into standard form by counting the zeros after the first digits.

   $10\,000 \times 100\,000 = 1.0 \times 10^4 \times 1.0 \times 10^5$

   Remember that you can add the powers when multiplying powers of 10.

   $1.0 \times 10^4 \times 1.0 \times 10^5 = 1.0 \times 10^9$

1 Using numbers and percentages

1 Using numbers and percentages   **11**

ii **100 and 0.00001**

Convert into standard form. For the small number, count the digits to the right of the decimal place and include the first significant figure:

$100 \times 0.00001 = 1.0 \times 10^2 \times 1.0 \times 10^{-5}$

Add the powers:

$1.0 \times 10^2 \times 1.0 \times 10^{-5} = 1.0 \times 10^{-3}$

Table 1.5 Unit prefixes

Prefix name	Symbol	Power of 10
tera	T	10^{12}
giga	G	10^9
mega	M	10^6
kilo	k	10^3
deci	d	10^{-1}
centi	c	10^{-2}
milli	m	10^{-3}
micro	µ	10^{-6}
nano	n	10^{-9}
pico	p	10^{-12}

Prefixes and converting units

The SI system of units has a set of prefixes that is used to avoid the power of 10 notations and thus simplify the way that units are presented. A familiar example is likely to be that of 1000 m, written as 1.0×10^3 m in standard form, which is 1 km. The common prefixes are given in Table 1.5 and you should learn them.

Converting units

In design and technology activities you will regularly need to convert the units of values in order to work with them or to make a calculation easier.

Consider the following example of an exam-style question.

Ⓐ Worked example

A flotation aid is used for teaching swimming.

■ **The volume of the thermopolymer foam is 0.00675 m³**
■ **The density of the thermopolymer is 0.05 g cm⁻³**
■ **mass = volume × density**

Calculate the mass, in kg, of the flotation aid.

This problem includes mixed units for:

■ length: cm and m
■ mass: g and kg.

To do the calculation, the units relating to length should all be converted to the common SI base unit of m, metres, and the units of mass should be converted to the common SI base unit of kg, kilograms.

Therefore, the density value needs to be converted into units of kg m⁻³ (see page 14).

A problem of this type, comprising mixed units, can be quite daunting and you need to be confident in the conversion of the units in order to get the correct answer. An answer could potentially be incorrect by many powers of 10. In a real-life situation, this could lead to wasted materials and expense.

The hardest units to convert are usually units of area and volume because these involve squaring and cubing, which result in some huge numbers.

Table 1.6 summarises common unit conversions for length, area and volume. You will find the standard form of the conversions useful in your calculations and you should learn them.

Table 1.6 Length, area and volume unit conversions

Units	Length in standard form, m	Area in standard form, m^2	Volume in standard form, m^3
m, metres	1 m	$1\,m^2$	$1\,m^3$
cm, centimetres	0.01 m $= 1.0 \times 10^{-2}\,m$	0.01 m $\times$ 0.01 m $= 0.0001\,m^2$ $= 1.0 \times 10^{-4}\,m^2$	0.01 m $\times$ 0.01 m $\times$ 0.01 m $= 0.000001\,m^3$ $= 1.0 \times 10^{-6}\,m^3$
mm, millimetres	0.001 m $= 1.0 \times 10^{-3}\,m$	0.001 m $\times$ 0.001 m $= 0.000001\,m^2$ $= 1.0 \times 10^{-6}\,m^2$	0.001 m $\times$ 0.001 m $\times$ 0.001 m $= 0.000000001\,m^3$ $= 1.0 \times 10^{-9}\,m^3$

 Worked examples

a **Convert the following values into standard form with the correct prefix.**

i **0.001 m**

Convert into standard form. Count three digits after the decimal point:

$0.001\,m = 1.0 \times 10^{-3}\,m$

Now replace the power of 10 with the correct prefix:

$1.0 \times 10^{-3}\,m = 1\,mm$

ii **240 000 000 mg**

Convert into standard form. Count all the digits after the first significant digit:

$240\,000\,000\,mg = 2.4 \times 10^8\,mg$

Convert the units. Replace the prefix 'milli' with the power of 10:

$2.4 \times 10^8\,mg = 2.4 \times 10^8 \times 10^{-3}\,g$

When multiplying, add the powers:

$2.4 \times 10^8 \times 10^{-3}\,g = 2.4 \times 10^5\,g$

Convert to the base unit kg, replace the power of 10 for the prefix kilo, i.e. 10^3:

$2.4 \times 10^5\,g = 2.4 \times 10^2\,kg$

iii **345 000 mm³**

Convert into standard form. Count the five digits after the first significant digit:

$345\,000\,mm^3 = 3.45 \times 10^5\,mm^3$

Convert mm^3 into m^3. Substitute mm^3 with $1.0 \times 10^{-9}\,m^3$:

$3.45 \times 10^5\,mm^3 = 3.45 \times 10^5 \times 10^{-9}\,m^3$

Add the powers:

$3.45 \times 10^5 \times 10^{-9}\,m^3 = 3.45 \times 10^{-4}\,m^3$

b Mass calculation

Here is the worked solution to the example question from the start of the section 'Converting units'.

A swimming flotation aid is used for teaching swimming.

- **The volume of the thermopolymer foam is 0.00675 m³**
- **The density of the thermopolymer is 0.05 g cm⁻³**

Calculate the mass, in kg, of the flotation aid.

First convert the volume into standard form to avoid errors:

$$0.00675\,\text{m}^3 = 6.75 \times 10^{-3}\,\text{m}^3$$

Next, for the density convert g to kg:

$$0.05\,\text{g cm}^{-3} = 0.00005\,\text{kg cm}^{-3} = 5.0 \times 10^{-5}\,\text{kg cm}^{-3}$$

Then for the density convert cm³ to m³:

$$5.0 \times 10^{-5}\,\text{kg cm}^{-3} = \frac{5.0 \times 10^{-5}}{1.0 \times 10^{-6}}\,\text{kg m}^{-3}$$

Dividing negative powers is the same as multiplying them, so for the case above:

$$\frac{5.0 \times 10^{-5}}{1.0 \times 10^{-6}} = 5.0 \times 10^{-5} \times 1.0 \times 10^{6}$$

Remember that when powers of the same number are multiplied, simply add the powers – in this case −5 and 6 become 1.

$$5.0 \times 10^{-5} \times 1.0 \times 10^{6} = 5.0 \times 10^{1}$$

Recall that any number to the power of 1 is the number multiplied by 1. In this case

$$10^1 = 10$$

$$5.0 \times 10^1 = 50\,\text{kg m}^{-3}$$

Finally, calculate the mass by using and rearranging the density formula from Section 3.4:

$$\text{density} = \frac{\text{mass}}{\text{volume}}$$

$$\rho = \frac{m}{V}$$

Rearranged into terms of m:

$$m = \rho V$$

Plugging in the values gives:

$$m = 5 \times 10^1 \times 6.75 \times 10^{-3} = 33.75 \times 10^{-2} = 3.375 \times 10^{-1}\,\text{kg}$$

The least accurate value that was provided was the density at 0.05 g cm⁻³, which has only one significant figure, so this result should be given to one significant figure.

$$m = 3.375 \times 10^{-1}\,\text{kg} = 3.0 \times 10^{-1}\,\text{kg or } 0.3\,\text{kg (1 s.f.)}$$

Answer: the mass of the swimming flotation aid is 0.3 kg (1 s.f.).

Accuracy

In design and technology, accuracy, or working accurately, is often critical to the success of a design.

When measuring, values should be given to the accuracy of the measuring equipment. For example, most digital calipers will measure to the nearest hundredth of a mm, 0.01 mm.

If all of the heights of your classmates were measured and you calculated the average height on your calculator, you might return a result such as 170.5678 cm. This value has four decimal places, which suggests that the original measurements were taken to this accuracy, which would require the use of expensive precision-measuring equipment. However, it is more likely that you would have measured your classmates to the nearest centimetre with a tape measure. Therefore, this result should be given to the same accuracy — it should be rounded up to 171 cm.

In the case of an exam question, the accuracy of the values provided at the start of the calculation should be used for your result, just as they were in the example above.

Rounding

Integers are **whole numbers**, such as 1, 2, 3, 4, not fractions.

Rounding makes a number shorter or simpler while keeping it close to the original value. The most common form of rounding is rounding to the nearest integer.

Rounding of numbers is quite simple. Just remember that if there is a **5 or higher** in the column of units to be rounded, **round up** to the nearest integer. If the units are **lower than 5, round down** to the nearest integer.

For example:

- 10.6 mm given to the nearest mm would round up to 11 mm because it is nearer to 11 mm than 10 mm.
- 137.2 N to the nearest newton rounds down to 137 N.

In both these cases it is the digits in the tenths place (first column to the right of the decimal point) that are being rounded. They are both fairly easy to round because there are no numbers in the hundredths column to confuse you. The next example is a little trickier.

The following numbers will all round down to 6 mm:

- 6.49 mm
- 6.4999 mm
- 6.499999 mm
- 6.4999999999999999999999999999999999999 mm

You may be tempted to round the '9s'. This would cause the number to be rounded up to 7 mm, which is wrong because for 6.4999999999999999999…, no matter how infinitesimally close to 6.5 it is, it is still less than 6.5 and the rule of rounding means that a number less than 6.5 should round down to 6.

The best way to tackle this is to avoid 'double rounding' and ignore all columns to the right of the tenths column. In the example, this means you will be looking at the .4 only, which, of course, rounds down, making the answer 6 mm.

Rounding is also used when answers are given to a certain level of significance.

Significant figures

The use of **significant figures** reflects the accuracy of a calculation. For example, if 1234 mm is given to 2 significant figures, it becomes 1200 mm, which is less accurate. When you write your answer, the words 'significant figures' can be abbreviated to 's.f.'. For example, '1200 mm to two significant figures' is written as 1200 mm (2 s.f.). This particular quantity should then be written in standard form — 1.2 m (2 s.f.). This tells the reader that the actual value is greater than or equal to 1.15 m and less than 1.25 m. Thus it is accurate to the nearest 0.1 m.

The use of standard form helps to identify the significant figures and the powers of ten. This is good practice and you should do it as it can prevent errors caused when working with large or very small numbers, particularly with lots of zeros. For example, 100 000 can easily be confused with 1 000 000. Similarly, 0.0000456 is easily confused with 0.00000456.

To help you decide on the number of significant figures you should use, remember the following: **zeros in front do not count; zeros behind do count.**

For example:

- 0.0000456 has three significant figures after the zeros. This should be written as 4.56×10^{-5}.

Decimal places

Most exam questions will identify the accuracy required for the final answer using either significant figures or decimal places. **'Decimal places'** is abbreviated to 'd.p.'. For example, '0.23 V to two decimal places' is written as 0.23 V (2 d.p.).

'Decimal places' always refers to the number of places to the right of the decimal point. For example, 12.34211 A given to an accuracy of 2 d.p. is 12.34 A (2 d.p.).

This is not always the same as the number of significant figures. For example:

- 0.12 A has two significant figures, and also two decimal places.

- 12.34 A has four significant figures, but two decimal places.

- 123.45 A has five significant figures and two decimal places.

(A) Worked examples

Rewrite the following numbers to 2 significant figures and the most appropriate prefix.

a **1573 mm**

Rewritten to 2 s.f., 1573 mm becomes 1600 mm because the 7 rounds up.

$1573 \text{ mm} = 1600 \text{ mm (2 s.f.)} = 1.6 \times 10^3 \text{ mm (2 s.f.)}$

Replacing milli with the equivalent power of 10 gives:

$1.6 \times 10^3 \text{ mm} = 1.6 \times 10^3 \times 10^{-3} \text{ m}$

Multiplying powers of ten is the same as adding the powers, which gives:

$$1.6 \times 10^3 \times 10^{-3}\,\text{m} = 1.6\,\text{m}$$

Answer: 1.6 m (2 s.f.)

b   0.0000567 A

Rewrite the value in standard form to 2 s.f. The 7 rounds up:

$$0.0000567\,\text{A} = 5.7 \times 10^{-5}\,\text{A}$$

Now choose the prefix milli = 10^{-3} and substitute this into the answer:

$$5.7 \times 10^{-5}\,\text{A} = 5.7 \times 10^{-2}\,\text{mA}$$

Answer: $5.7 \times 10^{-2}\,\text{mA}$ (2 s.f.)

c   17 433 g

Rewrite to 2 s.f. The 4 rounds down:

$$17\,433\,\text{g} = 17\,000\,\text{g} \ (2 \text{ s.f.})$$

Rewrite in standard form:

$$17\,000\,\text{g} = 1.7 \times 10^4\,\text{g}$$

Kilo would be a more suitable prefix and is the base unit. Kilo = 10^3 which gives:

$$1.7 \times 10^4\,\text{g} = 1.7 \times 10^1\,\text{kg} = 17\,\text{kg} \ (2 \text{ s.f.})$$

Answer: 17 kg (2 s.f.)

1.2 Working with formulae and equations

A **formula** is a rule describing the relationship between different quantities. For example, the formula for the area of a circle is:

$$\text{area of circle} = \frac{\text{circumference} \times \text{radius}}{2} = \frac{2\pi r \times r}{2} = \pi r^2$$

which gives the relationship between area and radius, r, of a circle.

An **expression** is a grouping of numbers, symbols and operators that represents a number or quantity, for example $3x$ or $2t$.

An **equation** is a mathematical statement that says that two expressions are equal. Equations always have an '=' symbol. For example:

$x + 3 = 7$

$x^2 = 25$

Substituting values

In design and technology, there are frequently calculations that involve formulae and it is important for you to be confident when **substituting values** into them. You will generally need to substitute values that you already have, or are given in an exam question, into the formula that you are using to complete the calculation.

Scientific formulae

Table 1.7 summarises the scientific formulae that you may already have learned from GCSE science and mathematics.

Table 1.7 Scientific and engineering formulae learned at GCSE

Formula in words	Formula in symbols	Notes
density $(\text{kg}\,\text{m}^{-3}) = \dfrac{\text{mass}\,(\text{kg})}{\text{volume}\,(\text{m}^3)}$	$\rho = \dfrac{m}{V}$	
pressure $(\text{N}\,\text{m}^{-2}) = \dfrac{\text{force}\,(\text{N})}{\text{area}\,(\text{m}^2)}$	$P = \dfrac{F}{A}$	
force (N) = mass (kg) × acceleration $(\text{m}\,\text{s}^{-2})$	$F = ma$	Newton's second law of motion
spring force (N) = spring constant $(\text{N}\,\text{m}^{-1})$ × extension (m)	$F = ke$	Hooke's law
weight (N) = mass (kg) × gravitational field strength $(\text{N}\,\text{kg}^{-1})$	$W = mg$	$g = 9.81\,\text{N}\,\text{kg}^{-1}$
potential difference (V) = current (A) × resistance (Ω)	$V = IR$	Ohm's law

SI units

Scientific and engineering formulae are a mathematician's way of showing relationships between physical quantities. Most quantities are measured in **SI units**. The physical quantities associated with scientific and engineering formulae, with their units and symbols, are listed in Table 1.8.

Table 1.8 Summary of SI units and symbols

Physical quantity	SI unit	SI unit symbol
mass	kilogram	kg
length	metre	m
time	second	s
force	newton	N
weight	newton	N
current	amp	A
potential difference	volt	V
resistance	ohm	Ω
temperature	degrees Celsius	°C
pressure	pascal	Pa (note: $1\,\text{Pa} = 1\,\text{N}\,\text{m}^{-2}$)
area		m^2
volume		m^3
density		$\text{kg}\,\text{m}^{-3}$
speed (velocity)		$\text{m}\,\text{s}^{-1}$
acceleration		$\text{m}\,\text{s}^{-2}$

TIP

Whenever you use a formula, be careful to put all the quantities into their proper SI units, otherwise the result may be incorrect.

A Worked examples

a **For the following formulae, substitute the value of radius *r* with 5 mm to calculate the answer.**

$$\text{area of a circle} = \frac{\text{circumference} \times \text{radius}}{2} = \frac{2\pi r \times r}{2} = \pi r^2 = \pi \times 5^2 = 78.5\,\text{mm}^2\,(3\,\text{s.f.})$$

$$\text{circumference of a circle} = 2\pi r = 2\pi \times 5 = 10\pi = 31.4\,\text{mm}\,(3\,\text{s.f.})$$

b **For the following formula, substitute the value of mass *m* with 10 kg and acceleration *a* with 9.8 m s^{-2} to calculate the answer.**

force = mass × acceleration

$$F = m \times a = 10 \times 9.8 = 98\,\text{N}$$

B Guided questions

Copy out the workings and complete the answers on a separate piece of paper.

1 **Calculate the surface area of a desktop that is 400 cm wide × 1.2 m long.**

area of a rectangle = width × length

Step 1: Convert the unit of the width into metres.

Step 2: Now multiply the width and length to calculate the area.

2 **Calculate the voltage difference across a resistor of 330 Ω when a current of 0.020 A flows through it.**

To calculate the voltage difference, you will need to use Ohm's law, which is:

voltage (volt, V) = current (ampere, A) × resistance (ohm, Ω)

$$V = IR$$

Step 1: Convert the current into standard form.

Step 2: Substitute the values into the formula.

Step 3: Rewrite this in standard form and present your answer to the correct level of accuracy.

3 **Calculate the volume of a cylinder that has a radius of 12 mm and a height of 0.2 m.**

volume of a cylinder = $\pi r^2 h$

where *r* is the radius and *h* is the height.

Step 1: Convert the radius into metres.

Step 2: Substitute the values into the formula.

Step 3: Rewrite this in standard form and present your answer to the correct level of accuracy.

C Practice questions

4 A laptop measures 360 mm long, 250 mm wide and 15 mm high. Calculate the area of the footprint, i.e. the area touching the desk.

5 A desk lamp has a cylindrical base. The base has a volume of $3.5 \times 10^{-3}\,\text{m}^3$. The base needs to have a minimum mass of 1.0 kg to prevent the lamp from toppling over. Calculate the density of material required for the base.

6   A green LED has a typical forward voltage of 2.1 V and an associated current of 20 mA. Calculate the power of the LED. Use the formula $P = IV$.

Rearranging formulae and equations

You will face many situations where you will need to **rearrange a formula** to make the subject of the formula the quantity that you require.

(A) Worked examples

a   **Rearrange Newton's second law so that it is in terms of a.**

   $F = ma$

   Step 1: Divide both sides of the equation by m:

   $$\frac{F}{m} = \frac{\cancel{m}a}{\cancel{m}}$$

   which becomes:

   $$\frac{F}{m} = a$$

   Step 2: Rewrite with terms in a on the left-hand side:

   $$a = \frac{F}{m}$$

b   **Pythagoras' theorem** is an important formula that defines the relationship between the sides of a right-angled triangle:

   $A^2 = B^2 + C^2$

   where A, B and C are the lengths of the sides and A is the longest side. Express the length of side B in terms of A and C.

   In order to calculate the length of side B, the formula needs to be rearranged into terms of B as follows.

   Step 1: Rearrange to leave terms of B on one side of the equation by subtracting C^2 from both sides.

   $$A^2 - C^2 = B^2 + C^2 - C^2$$

   This becomes:

   $$A^2 - C^2 = B^2$$

> **TIP**
>
> When rearranging formulae and equations, remember to do the same operation to both sides of the equals sign.

   which can be rewritten to have the terms of B on the left-hand side by simply swapping the expressions over as follows:

   $$B^2 = A^2 - C^2$$

   Step 2: In order to get A in terms of B, the square root of B^2 needs to be found. Again, apply the square root to both sides of the equation.

   $$\sqrt{B^2} = \sqrt{A^2 - C^2}$$

   This becomes:

   $$B = \sqrt{A^2 - C^2}$$

   Now values for A and C can be substituted to find B.

c The **voltage** or **potential** divider equation for the circuit diagram in Figure 1.2 is given as an equation of two ratios as follows:

$$\frac{V_{IN}}{V_{OUT}} = \frac{R1 + R2}{R2}$$

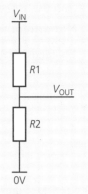

V_{IN}

$R1$

V_{OUT}

$R2$

0V

Figure 1.2
Voltage divider

The ratio of the voltages $V_{IN} : V_{OUT}$ is the same as the resistances $R1 + R2 : R2$

Rearrange the formula to find $R2$.

If the value of resistor $R2$ is needed, the equation needs to be rearranged to be in terms of $R2$. This is trickier because $R2$ is present in both numerator and denominator of the right-hand side of the equation.

Step 1: Multiply both sides by $R2$ in order to remove it from the denominator of the right-hand side.

$$\frac{V_{IN}}{V_{OUT}} \times R2 = \frac{R1 + R2}{\cancel{R2}} \times \cancel{R2}$$

which becomes:

$$\frac{V_{IN}}{V_{OUT}} \times R2 = R1 + R2$$

Step 2: Multiply both sides by V_{OUT} to simplify the equation.

$$\frac{V_{IN}}{\cancel{V_{OUT}}} \times \cancel{V_{OUT}} \times R2 = V_{OUT}(R1 + R2)$$

which becomes:

$$V_{IN}R2 = V_{OUT}(R1 + R2)$$

Multiply out $V_{OUT}(R1 + R2)$ in order to separate the terms in $R2$.

$$V_{IN}R2 = V_{OUT}R1 + V_{OUT}R2$$

Step 3: Arrange all terms in $R2$ on the left-hand side of the equation by subtracting $V_{OUT}R2$

$$V_{IN}R2 - V_{OUT}R2 = V_{OUT}R1 + V_{OUT}R2 - V_{OUT}R2$$

which becomes:

$$V_{IN}R2 - V_{OUT}R2 = V_{OUT}R1$$

Step 4: Factorise the left-hand side to separate $R2$.

$$R2(V_{IN} - V_{OUT}) = V_{OUT}R1$$

Step 5: Divide both sides of the equation by $V_{IN} - V_{OUT}$ to leave just terms in $R2$ on the left-hand side of the equation.

$$R2\frac{(\cancel{V_{IN}} - \cancel{V_{OUT}})}{\cancel{V_{IN}} - \cancel{V_{OUT}}} = \frac{V_{OUT}R1}{V_{IN} - V_{OUT}}$$

which becomes:

$$R2 = \frac{V_{OUT}R1}{V_{IN} - V_{OUT}}$$

d **Calculate the mass of an aluminium bar that has a volume of 0.0100 m³. The density of aluminium is 2.70 g cm⁻³.**

The density formula will be required for this calculation.

$$\text{density} = \frac{\text{mass}}{\text{volume}}$$

$$\rho = \frac{m}{v}$$

Step 1: Convert the density into base units of $kg\,m^{-3}$.

Remember that:

$$1\,g = 1.0 \times 10^{-3}\,kg$$

$$1\,cm^3 = 1.0 \times 10^{-6}\,m^3$$

Convert the units.

$$2.70\,g\,cm^{-3} = 2.70 \times 10^{-3}\,kg\,cm^{-3} = \frac{2.70 \times 10^{-3}}{1.0 \times 10^{-6}}\,kg\,m^{-3}$$

Remember that when dividing powers of 10 the powers can be subtracted. In this case $-3 - (-6) = -3 + 6 = 3$

Alternatively, dividing by a negative power is the same as multiplying by the positive power, so the formula can be rewritten as:

$$\frac{2.70 \times 10^{-3}}{1.0 \times 10^{-6}}\,\frac{kg}{m^3} = 2.70 \times 10^{-3} \times 1.0 \times 10^{6}\,kg\,m^{-3} = 2.70 \times 10^{3}\,kg\,m^{-3}$$

Step 2: Rearrange the density formula to make mass the subject by multiplying both sides by v.

$$\rho \times v = \frac{m}{\cancel{v}} \times \cancel{v}$$

which gives:

$$\rho v = m \qquad \text{or} \qquad m = \rho v$$

Step 3: Substitute values into the formula:

$$m = \rho v = 2.70 \times 10^{3}\,kg\,m^{-3} \times 0.0100\,m^3$$

$$= 0.0270 \times 10^{3}\,kg$$

$$= 27.0\,kg\ (3\ \text{s.f.})$$

The answer is given to 3 s.f. because this is the accuracy of both the values given at the start of the question.

Section 1.3 Equations of motion and 1.4 Scientific and engineering formulae are available online at www.hoddereducation.co.uk/essentialmathsanswers

2 Ratios and percentages

Ratios and percentages are both used to describe relationships between values in a simple form. In design and technology, ratios are commonly used for scaling drawings and percentages are used to describe changes, profit and waste. In this chapter, you will learn how to work with ratios and percentages in a range of contexts.

2.1 Ratios and scaling of lengths, area and volume

Ratios

A **ratio** is a relationship between two values that shows the number of times one value contains or is contained within the other. For example, widescreen televisions have a screen ratio of 16:9. This ratio relates to the width and height of the screen. It means that all widescreen televisions have a width that is $\frac{16}{9}$ of the height irrespective of their overall size.

$$\text{width} = \text{height} \times \frac{16}{9}$$

The width is a fraction of the height and the fraction is the ratio.

In design, the **golden ratio** of **1.618** is used because humans appreciate the aesthetics of this naturally occurring ratio. One side of the square in Figure 2.1 is multiplied by 1.618 to create a rectangle of harmonious proportions. The ratio of the height of the rectangle to the width of the rectangle is the ratio 1 : 1.618 because it has been **scaled** by 1.618. 1 mm on the side of the square scales up to become 1.618 mm on the rectangle, thus it has been multiplied by 1.618.

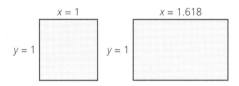

Figure 2.1 The golden ratio

Ratios of constituent parts

Ratios are often used to define the quantities of ingredients used. For example, when mixing concrete, a ratio is normally used. An example is 1 part cement, 3 parts sand and 3 parts aggregate. It doesn't matter what the unit of measurement is in this case; provided this ratio is used, when the water is used to bind the materials and the concrete is allowed to set, a standard result will be achieved.

 A Worked examples

In each case define the ratio of the parts.

a **Water, H_2O.**

A water molecule comprises two hydrogen atoms for every oxygen atom. Therefore, the ratio of hydrogen to oxygen is 2 : 1.

Answer: 2:1

b **A glue mix of 70% PVA glue, 30% water.**

The ratio of PVA to water is 70%:30%, which is the same as 7:3.

Answer: 7:3

c **A class comprising 12 girls and 13 boys.**

The ratio of girls to boys is 12:13.

Answer: 12:13

Scale ratios

You may be familiar with **scale ratios**. The scale ratio of a model, for example, represents how much smaller proportionally any linear dimension on the model is than the full-size product. A model car that is to a scale of 1:72 means that 1 mm on the model is 72 mm on the original. In the same way, if size is an issue you may need to make scale models and prototypes of your designs during your NEA project.

Similarly, it is normal for **plans and working drawings**, such as those created by architects and engineers, to be **drawn to scale** in order to enable very large objects to be represented on a sheet of paper.

Designers often work at a larger scale than that of the final product, or in other words 'larger than life'. This **'scaling-up'** allows designers to model and prototype small products in order to show detail clearly. Examples are jewellery fittings or the components of a ballpoint pen which might be drawn or modelled to a scale of 2:1 or 5:1. Microprocessors, such as the Intel Core i7, integrate more than 1 000 000 000 transistors and so these devices need to be scaled up enormously when they are designed.

Designers and manufacturers use **scaling ratios** when creating designs of different sizes. For example, the pattern for a size small t-shirt should be proportionally scaled up to create a pattern for a size medium t-shirt to ensure it increases in both length and width in the body as well as the sleeves. In the fashion industry, **pattern grading**, as this scaling is known, is often calculated using sizing charts and computer algorithms to more accurately reflect anthropometric data. In this section you will learn how to scale both two-dimensional shapes and three-dimensional objects, as well as calculate related dimensions from scaling operations.

Scaling ratios are normally written in the following form:

1:n

where n is the scaler.

However, in the case of decimal scalers, the scale ratio may be presented with integers (whole numbers), for example:

1:1.5 is the same scale ratio as 2:3 because 3 is 1.5 times larger than 2.

> **TIP**
>
> For the ratio: $x:y$
>
> If x is greater than y:
>
> $x > y$, the scale ratio reduces
>
> If x is smaller than y:
>
> $x < y$, the scale ratio increases

Scaling drawings and shapes

For any shape, the application of a scaling ratio will affect both width and height. For example, in Figure 2.2, the shape on the left has had a scaling ratio of 1 : 2 applied to it to create the shape on the right.

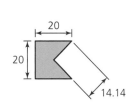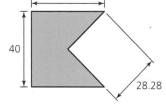

Figure 2.2 2D scaling

The inclusion of the dimensions of the diagonal line shows that it too has been scaled by the same ratio. In fact, all linear dimensions have been scaled by this ratio. This includes the perimeter of shapes.

This relationship is important as it can enable you to calculate dimensions from scaled shapes if you know the scaling ratio or if you can calculate the scaling ratio from the information provided. This does not apply only to two-dimensional shapes; it can also be applied to the side elevation or plan view of a three-dimensional object. In fact, for a three-dimensional object subjected to a scaling ratio, again, all linear dimensions will be scaled proportionally.

Scaling ratios can be represented as fractions in mathematical calculations. For example, the jug shown in Figure 2.3 has been scaled. The base of the jug is circular. To calculate the diameter x of the scaled down jug from the information provided, the scaling ratio needs to be calculated in order for the radius to be calculated.

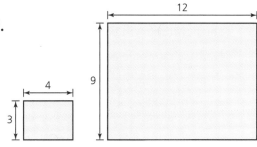

The scaling ratio for the diameters is the same as the height, therefore:

x : 100 mm is the same scaling ratio as 400 mm : 600 mm.

Figure 2.3 Scaled jug

Therefore the following relationship is true:

$$\frac{x}{100} = \frac{400}{600}$$

which gives x to be:

$$x = \frac{400}{600} \times 100 = 66.\dot{6} \text{ mm} = 66.7 \text{mm (3 s.f.)}$$

Note: The ˙ symbol above the 6 is called a recurring symbol and shows that the answer is a recurring '6', e.g. 66.6666666666666…

The values provided were accurate to 3 significant figures, so the value has been rounded to this accuracy.

Scaling area

For any scaling ratio 1 : n, the scaling ratio of the area is 1 : n^2.

This is because an area is a two-dimensional shape. If we think of this as having a width and a length, the area would be width × height. If both width and height have been scaled, then the product of the scaling ratios must be the scaling ratio multiplied by itself, in other words squared.

Figure 2.4 helps to prove this point.

Figure 2.4 Scaling area

The first rectangle has an area of $3 \times 4 = 12$.

The second rectangle has been subjected to a scaling ratio of $1:3$ and its area is $9 \times 12 = 108$.

The scaling ratio of the areas $= \dfrac{12}{108} = \dfrac{1}{9} = \dfrac{1}{3^2}$

$\qquad 1:3^2$

For ratios written with two integers greater than 1, for example $7:3$, both numbers are squared.

For any scaling ratio $x:y$, the scaling ratio of the area is $x^2:y^2$.

Scaling volume

For any scaling ratio $1:n$, the scaling ratio of the volume is $1:n^3$.

Volumes are three-dimensional in form, and just like the proof for area, if all three dimensions are multiplied by the same scaler, we can say that the scaler has been cubed. Again, this can be proven with a figure, as shown in Figure 2.5.

For the first cuboid:

$\qquad$ volume = width × length × height

$\qquad = 2\,\text{mm} \times 2\,\text{mm} \times 3\,\text{mm} = 12\,\text{mm}^3$

The second cuboid is subjected to a scaling ratio of $1:2$.

For this cuboid:

$\qquad$ volume = width × length × height

$\qquad = 4\,\text{mm} \times 4\,\text{mm} \times 6\,\text{mm} = 96\,\text{mm}^3$

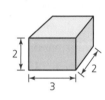 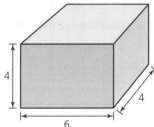

Figure 2.5 Scaling volume

The scaling ratio of the volumes $= \dfrac{12}{96} = \dfrac{1}{8} = \dfrac{1}{2^3}$

$\qquad 1:2^3$

As with area, for ratios written with two integers greater than 1, for example $7:3$, both numbers are cubed.

For any scaling ratio $x:y$, the scaling ratio of the volume is $x^3:y^3$.

(A) Worked examples

a **If the width of the large rectangle in Figure 2.4 is increased to 20 mm, what will the height be?**

The scale ratio of the large rectangle is:

$\qquad$ width : height

12 mm : 9 mm, which is the same as: $4:3$.

Or, in other words, the height is three-quarters of the width.

The new ratio will be: 20 mm : new height

As these ratios are equal:

$$\frac{20}{\text{new height}} = \frac{4}{3}$$

Rearranging gives:

$$\frac{\text{new height}}{20} = \frac{3}{4}$$

In terms of the new height this is:

$$\text{new height} = \frac{3}{4} \times 20 = 15\,\text{mm}$$

Answer: 15 mm (2 s.f.)

b What will the new area of the same rectangle be if it is subjected to a scale ratio of 5 : 2?

First notice that the scaling ratio is a reduction because the number on the right-hand side of the colon is smaller than the number on the left. This will remind you that you should end up with a result smaller than what you should at the current scale.

The current area is:

$$\text{area of rectangle} = \text{width} \times \text{height} = 12\,\text{mm} \times 9\,\text{mm} = 108\,\text{mm}^2$$

For any scaling ratio $1 : n$, the scaling ratio of the area is $1 : n^2$.

But remember that for ratios written with two integers greater than 1:

For any scaling ratio $x : y$, the scaling ratio of the area is $x^2 : y^2$.

This means that the area scale ratio is:

$$5^2 : 2^2$$

which is: $25 : 4$

Therefore, the new area is:

$$\text{new area} = \frac{4}{25} \times 108 = 17.28\,\text{mm}^2 = 17\,\text{mm}^2\,(2\ \text{s.f.})$$

You could check this by scaling the dimensions to calculate the area rather than calculating the area scale ratio.

Proof: Apply the scale ratio to each of the dimensions.

Applying the ratio 5 : 2 will create dimensions that are $\frac{2}{5}$ of the original, as proven in part (a) above.

$$\text{new width} = 12\,\text{mm} \times \frac{2}{5} = 4.8\,\text{mm}$$

$$\text{new height} = 9\,\text{mm} \times \frac{2}{5} = 3.6\,\text{mm}$$

$$\text{new area} = 4.8 \times 3.6 = 17.28 = 17\,\text{mm}^2\ (2\ \text{s.f.})$$

which gives the same answer.

Answer: The new area is 17 mm² (2 s.f.).

c What is the new volume of the large cuboid in Figure 2.5 if it is scaled by a ratio of 2 : 3?

This ratio should increase the volume because $3 > 2$.

$$\text{volume of original cuboid} = \text{length} \times \text{width} \times \text{height} = 4 \times 6 \times 4 = 96\,\text{mm}^3$$

For any scaling ratio $x : y$, the scaling ratio of the volume is $x^3 : y^3$.

Therefore, the volume scaling ratio is:

$2^3 : 3^3$

which is: $8 : 27$

Therefore, the new volume is:

new volume $= \dfrac{27}{8} \times$ original volume $= \dfrac{27}{8} \times 96 = 324 = 320\,\text{mm}^3$ (2 s.f.)

Answer: The new volume is $320\,\text{mm}^3$ (2 s.f.).

B Guided questions

1   The area of the large rectangle in Figure 2.4 is scaled up to 200 mm². Calculate the lengths of the new sides.

   The scale ratio for the sides can be calculated from the area scale ratio because for any scaling ratio $x : y$, the scaling ratio of the area is $x^2 : y^2$.

2   The large cuboid in Figure 2.5 is scaled to a volume of 150 mm³. What is the height of the new cuboid?

   As shown on p. 26, the original cuboid has a volume of 96 mm³. The height is 4 mm.

   Step 1: Use the volume scaling ratio to find the scaling ratio for the height.

   Step 2: Calculate the new height.

3   The large rectangle in Figure 2.4 is made of cotton fabric and costs 15 pence. A scaled-up version of the rectangle of fabric costs £2.10. What are its dimensions?

   For the rectangle, cost is proportional to the area. Therefore, the cost scaling ratio is the same as the area scaling ratio.

   Step 1: Use the cost scaling ratio to calculate the dimension scaling ratio.

   Step 2: Calculate the new width and height.

C Practice questions

4   If the large jug in Figure 2.3 is subjected to a scaling ratio of $2 : 3$, what will be the height of the new jug?

5   Calculate the volume scaling ratio for the jugs shown in Figure 2.3.

6   The mass of the large cuboid in Figure 2.5 is 0.4 kg. A cuboid with the exact proportions and of the same material has a mass of 1.4 kg. Calculate the height of the cuboid with the larger mass.

Section 2.2 Ratios and mechanisms is available online at www.hoddereducation.co.uk/essentialmathsanswers

2.3 Working with percentages

It is important to be confident in the use and application of percentages. In this section you will work through the various ways that percentage calculations are undertaken. This will help you respond to exam questions where percentage calculations may be applied to any data such as anthropometrics, survey results, measurements, profit and waste.

Calculating percentages on a calculator

Depending on the type of calculator you have, the % button may function in different ways.

On most calculators, the % button will convert a percentage number into a decimal, for example button presses **90 % =** will give the answer **0.9**.

On some calculators, the % function will apply percentages of numbers. For example, for button presses **90 + 10 % =**, the answer of **99** is given.

However, on some calculators the same button presses will result in the decimal conversion being added to the number, for example **90 + 10 % =** gives **90.1**.

> **TIP**
>
> If you intend to use the % button on your calculator, it is really important to determine how it functions for a range of calculations before relying on the answer.
>
> You need to be familiar with this function before using it for examination calculations. Be careful if you borrow a different model of calculator from a friend — try some test calculations, as it may function differently.

Converting fractions to percentages

In most cases percentages are used instead of a ratio, or a fraction, to compare two values.

For example, a 50 mm long bolt is $\dfrac{50\,\text{mm}}{60\,\text{mm}}$, or $\dfrac{5}{6}$, of the length of a 60 mm bolt.

'Per cent' means per 100, in other words a percentage is a fraction of 100. To convert from a fraction to a percentage you need to multiply by 100.

Therefore, the 50 mm bolt is $\dfrac{5}{6} \times 100 = 83.3\%$ (3 s.f.) of the length of the 60 mm bolt.

Percentage change

Percentage change describes how much the new value of a quantity has changed as a percentage of the original value.

$$\% \text{ change} = \frac{\text{new value} - \text{original value}}{\text{original value}} \times 100$$

Percentage profit

Calculations relating to profit are similar to those of percentage change.

$$\% \text{ profit} = \frac{\text{selling value} - \text{cost value}}{\text{cost value}} \times 100$$

Percentage error

Percentage error is a calculation that defines how much an error is of the expected value. The calculation again looks similar to those above.

If the **measured value is smaller** than the expected value, use the following formula:

$$\% \text{ error} = \frac{\text{expected value} - \text{measured value}}{\text{expected value}} \times 100$$

If the **measured value is larger** than the expected value, use the following formula:

$$\% \text{ error} = \frac{\text{measured value} - \text{expected value}}{\text{expected value}} \times 100$$

Percentage increase and decrease

When applying a percentage increase, use the following formula:

$$\text{new value} = \frac{100 + \text{percentage increase}}{100} \times \text{original value}$$

which rearranges to give:

$$\text{percentage increase} = \left(\frac{\text{new value}}{\text{original value}} \times 100 \right) - 100$$

For a percentage decrease calculation, subtract the percentage decrease from 100, as shown on the formula below:

$$\text{new value} = \frac{100 - \text{percentage decrease}}{100} \times \text{original value}$$

which rearranges to give:

$$\text{percentage decrease} = 100 - \frac{\text{new value}}{\text{original value}} \times 100$$

Converting percentages to decimals

Converting percentages to decimals can help you save time when calculating percentage increase or decreases.

For a percentage increase, multiply by 1 + the decimal, e.g. for a 20% increase, multiply by $1 + 0.2 = 1.2$.

For a percentage decrease, multiply by 1 − the decimal, e.g. for a 20% decrease multiply by $1 - 0.2 = 0.8$.

Percentage waste

Most manufacturers will seek to minimise waste. Calculations that identify percentage waste of a commodity are useful to help cost saving. This could relate to any commodity where costs are incurred. Time, energy and material are some examples.

$$\% \text{ waste} = \frac{\text{value including waste} - \text{value excluding waste}}{\text{value including waste}} \times 100$$

A Worked examples

a   The price of a hinge has increased from 79p to 99p. Calculate the percentage increase in price.

$$\% \text{ change} = \frac{\text{new value} - \text{original value}}{\text{original value}} \times 100$$

$$\% \text{ change} = \frac{99 - 79}{79} \times 100 = 25\% \text{ (2 s.f.)}$$

Answer: The percentage increase in price is 25% (2 s.f.).

b   A mild steel bar is ordered to be 250 mm long. When it is delivered it is measured to be 256 mm long. Calculate the percentage error.

In this example, the measured value is greater, so the calculation becomes:

$$\% \text{ change} = \frac{\text{measured value} - \text{expected value}}{\text{expected value}} \times 100 = \frac{256 - 250}{250} \times 100$$

$$= 2.4\%$$

Answer: The percentage error in the length of the mild steel bar is 2.4%.

c   An electric guitar case measuring 60 mm high is lined with polymer foam. When tested with a range of electric guitars it is found to be too tight. The manufacturer believes that a 10% increase in height will compensate for the polymer foam. Calculate the new height.

$$\text{new value} = \frac{100 + 10}{100} \times 60 \text{ mm} = 66 \text{ mm}$$

A 10% increase is the same as multiplying by 1.1.

Answer: The new height of the guitar case is 66 mm.

d   Last year a bicycle shop sold 1250 bicycles; 1212 bicycles were sold this year. Calculate the percentage drop in sales.

$$\text{percentage decrease} = 100 - \frac{\text{new value}}{\text{original value}} \times 100 = 100 - \frac{1212}{1250} \times 100 = 3\% \text{ (nearest \%)}$$

Answer: The percentage drop in sales was 3% to the nearest %.

e   A child's bicycle costs £67 to manufacture. The cost to retailers is £95 each. Calculate the percentage profit that the manufacturer makes on the cost price of each bicycle.

$$\% \text{ profit} = \frac{95 - 67}{67} \times 100 = 42\% \text{ (2 s.f.)}$$

Answer: 42% (2 s.f.) profit is made by the manufacturer.

f   A drilling process should take a factory worker 20 seconds per hole. However, on average the worker is taking 21 seconds. Calculate the percentage of wasted time.

$$\% \text{ waste} = \frac{21 - 20}{21} \times 100 = 4.8\% \text{ (2 s.f.)}$$

Answer: 4.8% of the time taken is wasted.

B Guided questions

1 **A linkage in a mechanism needs to be shortened from 210 mm long to 200 mm long. Calculate the percentage reduction in length.**

There are two options for answering this question, a longer method and a simpler calculation.

Option 1: A long-winded method is to use the formula for percentage decrease:

new value = ((100 − percentage decrease)/100) × original value

Option 2:

Step 1: Divide the new value by the original value.

Step 2: Subtract the decimal from 1.

Step 3: Multiply by 100.

2 **A circular table top, of radius 600 mm, will be made out of plywood. A square piece of plywood measuring 1.2 m × 1.2 m is bought to make the table top. Calculate the percentage of plywood that will be waste.**

The areas of the two shapes will be required to calculate the percentage waste.

Step 1: Calculate the area of the square that includes the waste.

Step 2: Calculate the area of the circle that excludes the waste.

Step 3: Substitute the values in the following formula and calculate the percentage waste.

$$\% \text{ waste} = \frac{\text{value including waste} - \text{value excluding waste}}{\text{value including waste}} \times 100$$

3 **A manufacturer ordered 1200 circuit boards from a factory. 0.6% of the circuit boards were found to be faulty. Calculate the number of circuit boards that were faulty.**

Think of how the percentage would be calculated. This percentage relates to the ratio of faulty circuit boards to the total number of circuit boards.

C Practice questions

4 Percentage waste calculation.

For a batch of backpacks, 1.32 m² of calico is used per backpack. The backpack actually requires only 1.21 m² of calico. Calculate the percentage waste to two decimal places.

5 Percentage profit calculation.

A manufacturer sells a desk lamp product to distributors at £12 per unit for a minimum order of 50 units. The manufacturer makes a profit of 35% on this price. For orders of over 100 units, the manufacturer wants to make £3 profit per lamp. Calculate the individual desk lamp selling price for orders of over 100 units to achieve this profit.

6 Percentage error calculation.

A student designs a concrete seating bench. They calculate that exactly 1.5 bags of ready-mixed concrete are required to make the bench. Each bag will provide $9.0 \times 10^{-3} \text{ m}^3$ of concrete at a cost of £6.29 per bag. Unfortunately, they make a 12% error in the calculations and more concrete is required. Calculate the volume of concrete required.

3 Calculating surface areas and volumes

In this section you will be given the properties of shapes before moving on to calculating surface areas and volumes of objects.

The surface area of an object is often calculated to determine the quantity and cost of a material. For example, the area of a room is calculated to determine the quantity and cost of carpet or laminate floor boards. Quantities of surface applications such as paints, stains, veneers or even vinyl wraps are defined by area. For example, it may be necessary to calculate the surface area of the legs of a table so that the correct amount of paint can be purchased. Paint manufacturers provide information on the paint can that informs you of the area of coverage. Your school, or a supplier, may charge you for materials used or needed in terms of surface area, for example the area of a table top made from a specific manufactured board such as 18 mm thick birch-faced plywood.

Volumes of objects are calculated to identify quantity and cost of material required. For example, if the base of a table lamp is cast out of concrete, you will need to know how much concrete you require. You will also need to know the cost of the concrete to pay for it. It may be that the weight of the base of the table lamp is also required in order to prove that it will not tip over. To calculate the weight, volume is also required.

3.1 Properties and areas of two-dimensional shapes

This section will explore the properties of polygons and circles to help you confidently calculate their areas. Classes and names of shapes that you may be familiar with from your maths lessons will be used to lead you into design and technology-style maths questions.

Polygons

Polygons are two-dimensional shapes. They are made of straight lines and the shape is 'closed', meaning all the lines connect up. Any shape with a curved line, e.g. a circle, is not a polygon.

Polygon comes from Greek. 'Poly' means 'many' and 'gon' means 'angle'.

Regular or irregular?

Regular polygons have sides of equal length and angles of equal size. The hardwood strips shown in Figure 3.1 have a square cross-section. This is a regular polygon because all sides are the same length and all internal angles are equal as they are 90°.

Figure 3.1 Hardwood strips with square sections

Table 3.1 gives the properties of some common regular polygons. You should learn these. Quick recall of these properties will help you calculate areas and volumes speedily and confidently.

Table 3.1 Properties of regular polygons

Name	Shape	Number of sides	Internal angle	Sum of internal angles
Equilateral triangle		3	60°	180°
Square		4	90°	360°
Pentagon		5	108°	540°
Hexagon		6	120°	720°

It is useful to know that regular polygons are made up of triangles (see Figure 3.2). This will help you calculate the area of regular polygons, such as hexagons, and the volume of regular polygonal prisms later in this section.

It is important to know the properties of **irregular** polygons as well. These include simple shapes such as rectangles, parallelograms and trapeziums.

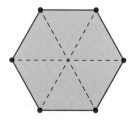

Figure 3.2 A polygon made of triangles

There is a useful formula to help you remember what the internal angles of a polygon add up to:

sum of angles = (*n* − 2) × 180°

where *n* is the number of sides of the polygon.

For example, the irregular polygon shown in Figure 3.3 has five sides. Therefore, the internal angles add up to:

$$(5 - 2) \times 180° = 3 \times 180° = 540°$$

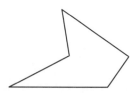

Figure 3.3 An irregular polygon

> **TIP**
>
> Alternatively, it is easy to remember that the internal angles of a triangle add up to 180°.

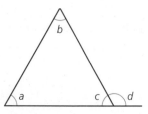

Figure 3.4 Internal angles of a triangle

In Figure 3.4 the angles *a*, *b* and *c* all add up to 180°. It is also worth noting that *c* and *d* add up to 180° — a useful fact.

A quick way to deduce the sum of the internal angles is to sketch the shape and see how many triangles you can fit into it, as shown Figure 3.5.

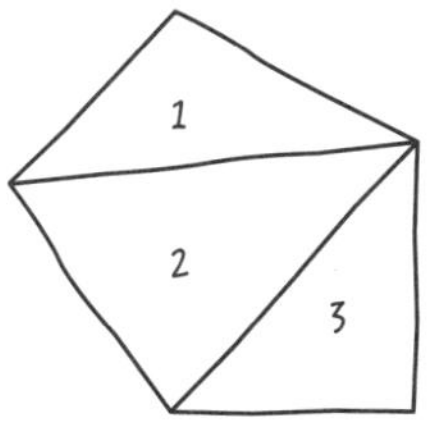

Now that you have a grasp of some of the basic concepts relating to shapes, you are ready to start calculating areas.

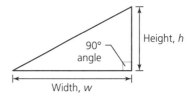

Figure 3.5 Freehand sketch of an irregular polygon

> **TIP**
>
> Area is a property of a two-dimensional shape. Calculating area involves the multiplication of two dimensions, for example width and height. It is important to remember that if you are given or need to calculate an area, the units will be squared, e.g. mm², cm² and m².

Squares and rectangles

Squares and rectangles both have right-angled corners. A right angle is 90°. Their area is calculated by multiplying the width of the base of the shape by the height.

area of a rectangle = width × height

> **TIP**
>
> The four right angles inside a square or rectangle add up to 360°. This is the same for all four-sided shapes, also known as quadrilaterals.

Triangles

Triangles can be classified as right-angled, equilateral, isosceles, acute and obtuse.

> **TIP**
>
> The three angles inside a triangle add up to 180°.

When calculating the area of a triangle it is important to consider the form of a triangle carefully to ensure that the correct method of calculation is chosen.

A **right-angled triangle** (see Figure 3.6) has one angle that is a right angle, or 90°. It fits into a rectangle of area = width, w × height, h.

Two of these triangles fit, or tessellate, into the rectangle, as shown in Figure 3.7. Therefore, the area of a right-angled triangle is half the area of the rectangle:

area of a right-angled triangle = $\frac{1}{2}$ × width, w × height, h

> **TIP**
>
> **Tessellate** means to cover a surface by repeated use of a single shape, without gaps or overlapping.

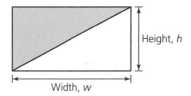

Figure 3.6 A right-angled triangle

Figure 3.7 Two identical right-angled triangles fit in a rectangle

An **isosceles triangle** (see Figure 3.8) is made up of two symmetrical right-angled triangles. Therefore, the two longest sides, known as hypotenuses, are the same length. The angles in the bottom corners are also symmetrical and the same.

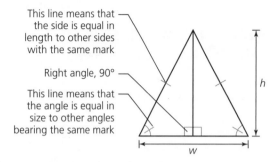

This line means that the side is equal in length to other sides with the same mark

Right angle, 90°

This line means that the angle is equal in size to other angles bearing the same mark

Figure 3.8 An isosceles triangle

Four of these identical triangles fit, or tessellate, into a rectangle of area = width, w × height, h, as shown in Figure 3.9.

Therefore, the area of an isosceles triangle is also:

$$\text{area of an isosceles triangle} = \frac{1}{2} \text{ width}, w \times \text{height}, h$$

The triangle shown in Figure 3.10 is a regular isosceles triangle that is a polygon known as an **equilateral triangle**.

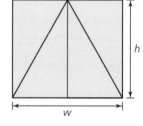

Figure 3.9 Rectangle contains four right-angled triangles

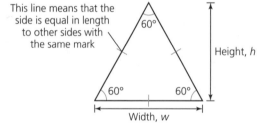

This line means that the side is equal in length to other sides with the same mark

60°

60°   60°

Height, h

Width, w

Figure 3.10 An equilateral triangle

From Table 3.1:

- all angles are equal, i.e. $\frac{180°}{3} = 60°$
- all sides are equal in length.

The area calculation is the same as for an isosceles triangle.

An **acute triangle**, as shown in Figure 3.11, has angles that are all less than 90°.

An **obtuse triangle**, as shown in Figure 3.12, has one angle that is greater than 90°.

When calculating the area of a triangle that has either an acute or an obtuse angle, a similar looking calculation can be made for the area, but with these triangles use the perpendicular, or vertical, height identified in Figures 3.11 and 3.12.

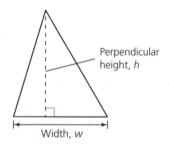

Perpendicular height, h

Width, w

Figure 3.11 An acute triangle

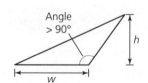

Angle > 90°

h

w

Figure 3.12 An obtuse triangle

$$\text{area of an acute or obtuse triangle} = \frac{1}{2} \times \text{width}, w \times \text{height}, h$$

To understand why this is the case it is important to be aware of the properties of a parallelogram.

Parallelograms

A parallelogram is an irregular polygon — a four-sided shape with parallel sides. Opposite angles are equal, as shown in Figure 3.13, where the obtuse triangle from Figure 13.12 is repeated to create a parallelogram.

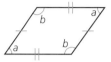

Figure 3.13 A parallelogram

The area of the parallelogram can be rearranged into the shape of a rectangle, as shown in Figure 3.14. This proves that the calculation for the area of a parallelogram is similar to that of a rectangle:

area of a parallelogram = width, w × height, h

This also proves that the area of the obtuse triangle must be half that of the parallelogram as given above.

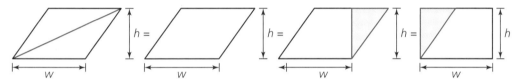

Figure 3.14 Parallelograms are made up of two identical obtuse triangles

Rhombus

A **rhombus**, or **diamond** shape, is actually a parallelogram. In a parallelogram, each pair of opposite sides is equal in length. With a rhombus, all four sides are the same length. It therefore has all the properties of a parallelogram and the area can be calculated in the same way by rotating the rhombus onto a side horizontally, as shown in Figure 3.15.

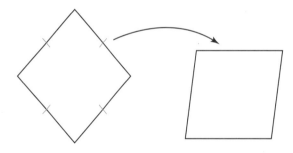

Figure 3.15 A rhombus is a parallelogram

Trapeziums

The key difference between a **trapezium** and a parallelogram is that one pair of sides is not parallel. This means that the area calculation is a little different and looks more like that of a triangle. For the trapezium in Figure 3.16, with sides of length a and b and perpendicular height h:

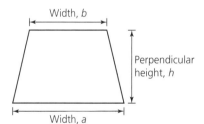

Figure 3.16 A trapezium

$$\text{area of trapezium} = \frac{\text{sum of parallel sides} \times \text{perpendicular height}}{2} = \frac{a+b}{2}h$$

Figure 3.17 helps to show why this is the case. The trapezium can be divided into a rectangle and two triangles.

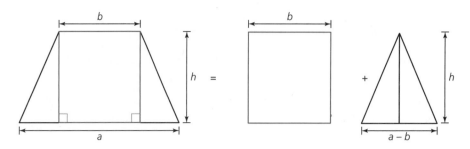

Figure 3.17 A trapezium comprises a rectangle and two triangles

It is now much more obvious how the area of the trapezium can be calculated. The process is simpler if the two triangles are joined, as shown, to form a larger triangle. Therefore, the area of the trapezium is:

area of trapezium = area of rectangle + area of triangle

$$= bh + \frac{1}{2}(a-b)h$$

$$= bh + \frac{ah}{2} - \frac{bh}{2}$$

$$= \frac{bh}{2} + \frac{ah}{2}$$

$$= \frac{a+b}{2}h$$

Pentagons and hexagons

When a pentagon or hexagon is mentioned, we generally think of the regular form of these shapes, even though it is possible to have irregular pentagons and hexagons. One method of calculating their area, and that of any other regular polygon, is to break up the shape into triangles, as shown in Figure 3.18.

The resulting triangles are isosceles triangles.

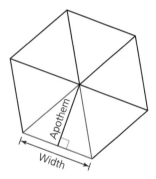

Figure 3.18 Triangles in a regular polygon

- The 'width' of the triangle is one side of the polygon.

- The 'height' of the triangle is known as the 'apothem' of the polygon.

The area of one of these triangles can be given as:

$$\text{area of isosceles triangle} = \frac{\text{width} \times \text{apothem}}{2}$$

The area of the whole regular polygon can be found by adding up the areas of the triangles, which gives:

$$\text{area of regular polygon} = n \times \frac{\text{width} \times \text{apothem}}{2}$$

Since the perimeter of all the sides $= n \times$ width

$$\text{area of regular polygon} = \frac{\text{perimeter} \times \text{apothem}}{2}$$

In Section 4.2 you will learn how to calculate the length of an apothem using trigonometry.

Circles

Circles are single curved-sided shapes, as are ovals. Therefore, they are not polygons. The perimeter of a circle is known as the **circumference** (see Figure 3.19).

For any circle, there is a special ratio known as **pi**, π.

$$\pi = \frac{\text{circumference}}{\text{diameter}}$$

Since the diameter is twice the length of the radius of the circle, this becomes:

$$\pi = \frac{\text{circumference}}{2 \times \text{radius}}$$

By rearranging the ratio, the following important formula can be found:

$$\text{circumference} = 2\pi r$$

where r is the radius of the circle, as shown in Figure 3.19.

Calculators have a π button and most will display the value correct to a minimum of seven significant figures as 3.141593.

The area of a circle can be found using:

$$\text{area of circle} = \frac{\text{circumference} \times \text{radius}}{2} = \frac{2\pi r \times r}{2} = \pi r^2$$

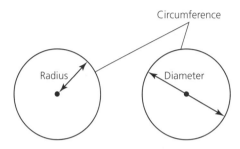

Figure 3.19 A circle

> **TIP**
>
> Try to remember the value of π to 3 significant figures: 3.14. It is a useful value that will enable you to make quick estimates.
>
> Remember: $\pi \approx 3.14$

Ellipses

An ellipse is a type of oval that is symmetrical in two perpendicular axes, as shown in Figure 3.20.

$$\text{area of ellipse} = \pi \times a \times b = \pi ab$$

where:

- a is known as the semi-major axis, which is the larger radius
- b is known as the semi-minor axis, which is the smaller radius.

The circle is actually a special case of an ellipse where both a and b are the same length of radius, r. If r is substituted into the area formula it becomes $\pi \times r \times r = \pi r^2$, which is correct.

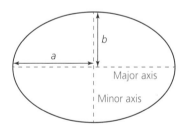

Figure 3.20 Ellipse

A Worked examples

a Length of a curved hanging rail

A curved hanging rail has been designed to be made from a tube of mild steel that is 25 mm in diameter. Use the dimensions given in Figure 3.21 to help you calculate the length of tube required for one hanging rail. Give your answer to the nearest mm.

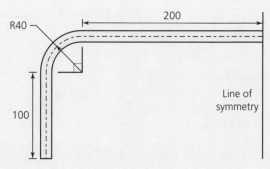

Figure 3.21 Hanging rail dimensions

It is important to remember that when a tube is bent, the outer wall elongates and the inner wall compresses. Therefore, it is necessary only to consider the length of the centre of the tube, which is identified by the dashed centre line. You do not need the diameter to answer this question.

The curve is a constant radius for 90° and thus one-quarter of the circumference of a circle. Pi can be used to calculate this.

$$\frac{\text{circumference}}{4} = \frac{2\pi r}{4} = \frac{\pi r}{2} = \pi \times \frac{40}{2} = 62.83 \, \text{mm to 2 d.p.}$$

The drawing shows a line of symmetry indicating that twice the length of the drawn part is required.

total length = (2 × 100) + (2 × 62.83) + (2 × 200) = 725.66 = 726 mm to the nearest mm.

Answer: The total length of mild steel tube required is 726 mm.

b Manufactured boards

Manufactured boards such as MDF and plywood are bought from suppliers in standard sizes of 8′ × 4′, or 2440 mm × 1220 mm. Calculate the surface area of the top face of a standard-sized manufactured board.

Area of a rectangle = width × height = 2.44 × 1.22 m = 2.9768 m² = 2.98 m² to 3 significant figures.

Answer: The area of the top face is 2.98 m² to 3 s.f.

This is a useful value to know. This value will be used later in Section 4.2.

c Laser-cutting right-angled triangles

A batch of triangular-shaped pieces of cotton is required for applique onto a batch of products. Each triangle should be right-angled, 150 mm wide and 200 mm high. The manufacturer buys cotton in rolls that are 900 mm wide.

Calculate the length of the roll of cotton that is required for 48 of the triangles to be laser-cut without waste.

For the purpose of this calculation you can ignore any waste caused by the cutting process of the laser cutter.

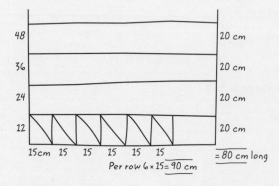

> **TIP**
>
> This is a tessellation question. A sketch like the one shown in Figure 3.22 will help you answer a question like this quickly.

Figure 3.22 A sketch helps to work out tessellation

First, identify whether the width or the height of the right-angled triangle are factors of the width of the roll. In this case, the 150 mm width of the triangle is a factor of the 90 cm width of the roll because 6 × 150 = 900.

Therefore, six of the triangles fit perfectly into the roll width ways, with no waste. As the triangles are right-angled, 12 will tessellate into rectangles in each 200 mm long strip of the roll of cotton.

48 of the triangles are needed: 48 ÷ 12 = 4 strips that are 200 mm long each.

Answer: 800 mm of the cotton roll is needed for 48 of the triangles.

B Guided questions

1 Desk area

A client would like a desk to be made to fit into the corner of a room. The edges should be parallel to the walls, as shown in the floorplan in Figure 3.23. The desktop should be 600 mm wide and 1200 mm long. The client would like to know what the surface area of the desktop will be. Calculate the total surface area that the desktop will provide for the client. Give your answer in units of m².

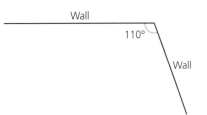

Figure 3.23 Floorplan

The desktop will be parallelogram shaped. This question could fool you into thinking it is necessary to use trigonometry to calculate the area. However, it is actually quite a simple question.

2 Percentage of waste material

Figure 3.24 is a drawing of the top panel of a wooden stool. A manufacturer wishes to modify the design of the stool top to create a hand hold for transportation. Calculate the waste material as a percentage of the wood used to make the top panel. Give your answer to the nearest whole percentage.

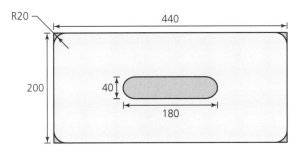

Figure 3.24 Stool top drawing

Step 1: The waste material is shown by the green area. In order to calculate a percentage, compare the area of the waste material to that of the original plank.

Step 2: Calculate the area of the waste. First, sketch over the drawing to identify the shapes of the waste material.

Step 3: As shown in Figure 3.25, the waste material can be simplified to reduce the number of calculations necessary. The four corners will make one square containing a circle and the two semi-circles will make a circle.

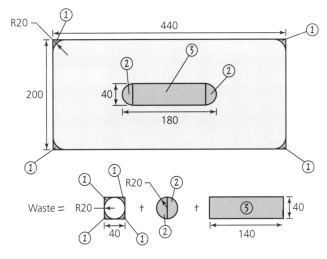

Figure 3.25 Edit of drawing

ⓒ Practice questions

3 Coffee table

Figure 3.26 shows the design of a coffee table made from laminated plywood. The table will be made by laminating eight layers of 1.5 mm flexible birch plywood around a former in a vacuum bag. The former is shown in Figure 3.27.

Figure 3.26 Coffee table

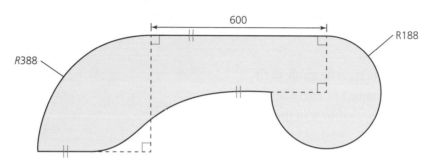

Figure 3.27 Side elevation of former

Calculate the length of the outside layer of plywood used to make the table. For the purpose of this calculation, ignore the additional thickness of adhesive used in the lamination process.

4 CNC routing speed question

A dining table has legs made from 19 mm plywood. The leg is cut out of a sheet of plywood by a CNC router. The 1200 mm long side (see Figure 3.28) takes three passes and 6 minutes in total to machine.

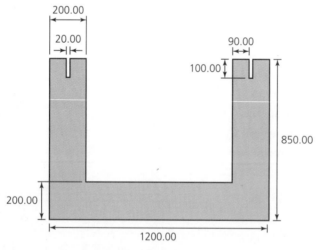

Figure 3.28 Dining table end leg

> **TIP**
>
> Annotate the drawing with the missing dimensions to help you.

For the purpose of the following calculations, ignore that the router would create a curved fillet on the inside corners.

a Calculate the CNC router cutting speed.

b Calculate the length of time it will take to cut the complete table leg out of the plywood sheet.

3.2 Surface area of three-dimensional objects

Since design and technology tasks involve designing and making three-dimensional objects, as well as their packaging, it is useful to know how to identify the two-dimensional shapes that make up the surfaces of these objects. This is a particularly important skill used when designing nets or developments for packaging, creating lay plans or designing with CAD. Areas are required to calculate volumes, so you will find that many of the questions in Section 3.3 also requires you to understand and know how to calculate the areas of different shapes.

Prisms

Prisms are three-dimensional objects that have the same cross-section throughout their length. In design and technology, many stock materials are available as prisms. For example, steel and aluminium extrusions are available in a range of cross-sections that includes circular (rods) and rectangular (bars). Tubes and pipes are available in circular, rectangular and flat-sided oval cross-sections.

You may also be familiar with a range of forms of acrylic and softwood stock materials that are prisms.

In this section the following examples will be considered:

- rectangular prisms known as cuboids

- triangular prisms

- circular prisms known as **cylinders**.

Figure 3.29 Prisms of a range of cross-sections found in the design and technology workshop

From working with these examples and your knowledge of more complex shapes, you should be able to work with prisms of a range of cross-sections familiar in the design and technology workshop (see Figure 3.29). These include tubes and rods of various shapes:

- semi-circular   ■ oval   ■ flat-sided oval (fso)   ■ hexagonal   ■ angle.

Cuboids

Cuboids, or boxes, have surfaces that are made up of squares and rectangles. The cuboid in Figure 3.30 is shown in three dimensions and a net has been drawn to show what the two-dimensional surfaces look like. In mathematics, a **net** is the shape that is formed by unfolding a three-dimensional figure. In design and technology this is known as a surface development, or just **development**.

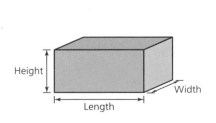

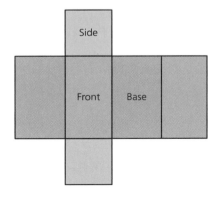

Figure 3.30 A cuboid and development

Their surface areas are simple to calculate: just add up the areas of the faces.

surface area of a cuboid = 2 × (area of base + area of one side + area of front)

Triangular prisms

Triangular prisms have surfaces that are made up of rectangles and triangles, as shown by the development in Figure 3.31.

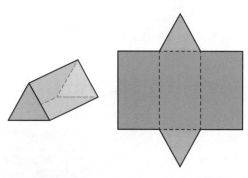

Figure 3.31 A triangular prism and development

surface area of a triangular prism = total area of all the rectangular faces +
(2 × area of one of the triangular ends)

For a triangular prism that has a cross-section that is an **equilateral triangle**, all of the rectangles will be identical, therefore:

surface area of an equilateral triangular prism = (3 × area of a rectangular side) +
(2 × area of one of the triangular ends)

Cylinders

Cylinders have circles at each end. The curved surface is actually rectangular — imagine a rectangle wrapped around the cylinder (see Figure 3.32).

The dimensions, height and width, of the rectangle can be given as:

height = height of the cylinder = h

width = perimeter, or circumference, of the circle = $2\pi r$

By applying the area of a rectangle formula from Section 3.1, the area of the curved surface of a cylinder is:

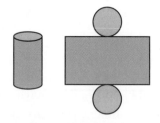

Figure 3.32 A cylinder and development

surface area of the curved surface of a cylinder =
circumference of base × height = $2\pi rh$, or πdh

where d is the diameter of the circle.

A cylinder also comprises a circle at each end, so:

surface area of a cylinder = $2\pi rh + 2\pi r^2$

Pyramids

Pyramids are traditionally considered to have regular square bases and a central vertex (see Figure 3.33). However, a pyramid can actually have a base the shape of any polygon,

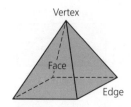

Figure 3.33 A regular pyramid

regular or not. The method to calculate surface area is the same for any of these pyramids.

For **regular pyramids**, such as the one shown in Figure 3.33, or any pyramid with a base the shape of a regular polygon and a central vertex, all of the triangular faces have the same area (see Figure 3.34):

surface area of any regular pyramid = area of base + (n × area of one of the triangular faces)

where n is the number of sides on the regular polygonal base, e.g. for the square-based pyramid in Figure 3.33, the calculation becomes:

surface area of pyramid = area of square base + (4 × area of one of the triangular faces)

In design and technology, designs are more likely to include **irregular pyramids** which have bases that are irregular polygons (see Figure 3.35). Vertices are more likely to be off-centre, such as the pyramid shown in the figure, and the triangular sides will have different areas:

surface area of any irregular pyramid = area of base + (sum of the areas of the triangular faces)

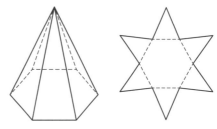

Figure 3.34 A regular pyramid and development

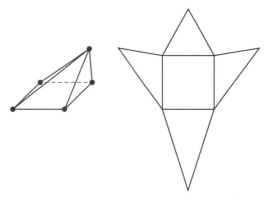

Figure 3.35 An irregular pyramid and development

Cones

A **cone** has a circle at the base. When the curved surface is considered in two dimensions, like a development, it is actually a sector of a large circle wrapped around the perimeter of the smaller base circle. This can be seen in Figure 3.36 — imagine the sector wrapped around the base to create the cone.

The dimensions, height and width, of the sector can be given as:

height = slant height of the cone = l

width = perimeter, or circumference, of the base circle = $2\pi r$

The formula for the area of a circle can be used to give the area of the curved surface of a cone as:

surface area of a cone = circumference × radius/2

$$= \frac{2\pi r l}{2} = \pi r l$$

area = base × height^2

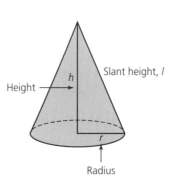

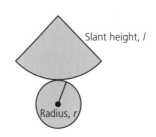

Figure 3.36 A cone

A **truncated cone** is a cone that looks like the top has been cut off. These are often seen in products such as the lamp counter-balance at the end of this section. In Figure 3.37 a truncated cone is shown. This truncated cone has had the top cut off parallel to the base of the cone.

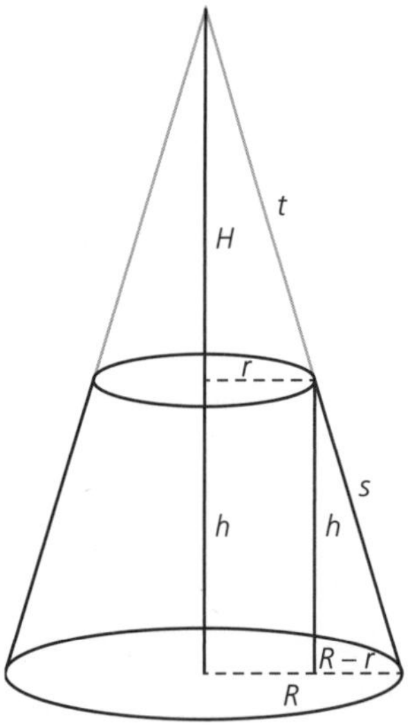

This type of **parallel truncated cone** is essentially a large cone with a smaller one removed from the top. In the case of Figure 3.37, the orange cone is the smaller cone to be removed. You could calculate the surface area of the larger cone and subtract the surface area of the smaller cone if you are given sufficient information.

$$\text{surface area of parallel truncated cone} = \pi R(s+t) - \pi rt$$

If not, it is a bit trickier to work with.

If you are not given the slant height, it can be found using Pythagoras' theorem, which is explained in Section 4.1, and gives:

$$\text{slant height, } S = \sqrt{(R-r)^2 + h^2}$$

The surface area of the curved surface or lateral surface can be found using:

$$\text{curved surface area of parallel truncated cone} = \pi S(r + R)$$

Figure 3.37 A truncated cone

Spheres

The surface area of a **sphere** (see Figure 3.38) is given as:

$$\text{surface area of a sphere} = 4\pi r^2$$

The curved surface area of the **hemisphere** (Figure 3.39) is half of this:

$$\text{curved surface area of a hemisphere} = 2\pi r^2$$

Figure 3.38 A sphere

Figure 3.39 A hemisphere

Ⓐ Worked examples

a   **Look back at Figure 3.21. The manufacturer would like to sell a version of the hanging rail that is chrome electroplated. Calculate the surface area of the outer curved face of the tube to the nearest mm².**

This looks like a tricky question as the tube is curved; however, remember that during the bending process, the inside of the tube is compressed and the outside is elongated. You can assume that the resulting surface area is the same as before bending. The length of the tube from Section 3.1 was 726 mm. The diameter of the tube was given as 25 mm. For surface area, the tube can be treated as a solid cylinder and only the curved-face surface is considered in this particular case.

$$\text{surface area of a cylinder} = 2\pi rh = 2\pi \times 12.5 \times 726$$
$$= 57\,020 \text{ mm}^2 \text{ to the nearest mm}^2$$

Answer: The surface area that will be electroplated is 57 020 mm² to the nearest mm².

b **A luxury chocolate manufacturer is considering packaging a product in a square-based pyramid-shaped container, as in Figure 3.40. Calculate the surface area of the packaging in mm².**

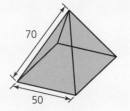

Figure 3.40 A luxury chocolate box

Remember, the net of a pyramid looks like Figure 3.34 and comprises a square base and four isoceles triangles. Also, from above:

surface area of any pyramid = area of base + (n × area of one of the triangular faces)

area of base = 50 mm × 50 mm = 2500 mm²

Next, calculate the height of one of the triangular faces using Pythagoras' theorem from Section 4.1 to calculate the height, h. Remember, these are isosceles triangles.

$$\text{height, } h = \sqrt{\text{slant height}^2 - \left(\frac{\text{base}}{2}\right)^2} = \sqrt{70^2 - \left(\frac{50}{2}\right)^2} = 65 \text{ mm}$$

Use the area of a triangle formula from Section 3.1.

$$\text{area of a triangle} = \frac{\text{width, } w \times \text{perpendicular height,} h}{2} = \frac{50 \text{ mm} \times 65 \text{ mm}}{2}$$
$$= 1625 \text{ mm}^2$$

surface area of the pyramid = 2500 + (4 × 1625) = 9000 mm²

Answer: The surface area of the packaging is 9000 mm².

B Guided question

1 **Surface coating of a ramekin**

The stainless steel ramekin shown in Figure 3.41 has a blue enamel paint coating to the outer curved surfaces and flat circular base. Calculate the surface area coated with blue enamel paint.

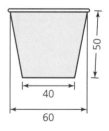

Figure 3.41 Stainless steel ramekin

For the curved surface, use the following formula:

curved surface area of parallel truncated cone = $\pi S (r + R)$

where:

S is the slant height 50 mm

r is the base radius

R is the larger radius at the top of the cone.

In addition, you will need to calculate the surface area of the circular base.

C Practice questions

2 Desk lampshade waste material

A desk lamp has a thin, mild steel, pressed shade that is hemispherical at one end of a cylinder. Figure 3.42b is a drawing of the side elevation of the lampshade.

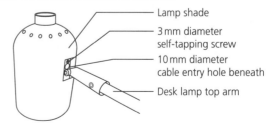

Figure 3.42a View from rear of lamp

In addition, the shade has:

- 12 ventilation holes, each 6 mm in diameter
- one 15 mm hole for the lamp holder
- two 3 mm diameter holes for the screws that attach the top arm
- one 10 mm diameter hole for the cable entry into the lampshade.

Calculate the waste material as a percentage of the mild steel sheet used to make the lampshade.

3 Surface paint application

The stool top shown in Figure 3.24 is going to be painted red before attaching it to the legs. The whole part needs to be painted. The top is 20 mm thick.

Calculate the surface area for a single coat of paint to cover. Give your answer in m² to 2 d.p.

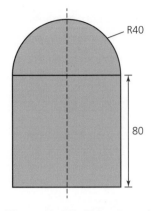

Figure 3.42b Side elevation of the lampshade

3.3 Volume of three-dimensional objects

Once you are confident calculating surface areas of objects it is easier to calculate their volumes. In this section the relationship between areas and volumes will be identified to make it easier for you to understand how the volumes are calculated.

> **TIP**
>
> Volumes are three-dimensional and the calculation of a volume will involve the multiplication of three dimensions, e.g. length × width × height. Units are always cubed, for example mm³, cm³ and m³.

Prisms

Calculating the volume of **prisms** is relatively straightforward because the shape of their cross-section is the same throughout their length. Therefore, for any prism:

volume of a prism = area of cross-section × height (or length)

The formulae for volume of popular prisms are given below.

Cuboids

For a **cuboid** (see Figure 3.43), the area of the cross-section is length × width, therefore:

volume of a cuboid = length × width × height = lwh

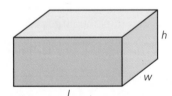

Figure 3.43 A cuboid

Cylinders

For a **cylinder** (see Figure 3.44), the area of the cross-section is πr^2, therefore:

volume of a cylinder $= \pi r^2 h$

r = radius
h = height

Figure 3.44 A cylinder

Pyramids

volume of a pyramid $= \dfrac{1}{3}$ base area $\times$ height

Pyramids may be triangle-based or square-based (Figure 3.45). The volume formula works for all types of pyramid.

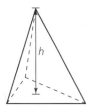 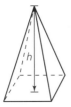

Figure 3.45 Triangle-based and square-based pyramids

> **TIP**
>
> For every pyramid, the volume is exactly one-third of the volume of a prism with the same base and height.

Cones

Remember, a cone is just a special pyramid, so the same formula is used to calculate volume. For the cone shown in Figure 3.36 the volume is:

volume of a cone $= \dfrac{1}{3}$ base area $\times$ height $= \dfrac{1}{3}\pi r^2 h$

Truncated cones

If you are given sufficient information, the volume of a truncated cone can be calculated by subtracting the volume of the small cone removed from the top from the volume of the complete cone.

For the truncated cone shown in Figure 3.37, the volume is:

volume of a truncated cone = volume of large cone − volume of small cone

$$= \frac{1}{3}\pi R^2(h + H) - \frac{1}{3}\pi r^2 H = \frac{1}{3}\pi(R^2(h + H) - r^2 H)$$

In most cases, you will not know what the height, H, will be as it will not exist, therefore it is necessary to define H in terms of dimensions that are known. From Figure 3.37 it can be seen that r, H and R, $h + H$ are related and the ratio of R to r is the same as the ratio of h to $h + H$.

This gives the relationship:

$$\frac{R}{r} = \frac{h + H}{H}$$

Rearranged in terms of H, this gives:

$$H = \frac{hr}{(R - r)}$$

When this is plugged into the formula, and after a bit of factorisation, it will become:

$$\text{volume of a truncated cone} = \frac{1}{3}\pi h(Rr + R^2 + r^2)$$

Note: The full working of this formula is not shown here as it is quite lengthy; however, it is readily available from reputable internet-based mathematics resources should you wish to explore it further.

Spheres

The approach is the same for hemispheres. The volume for the hemisphere in Figure 3.39 is therefore:

$$\text{volume of a hemisphere} = \frac{1}{3} \text{ curved surface area} \times \text{height} = \frac{2}{3}\pi r^3$$

In this case, the height is the radius, r.

The sphere is twice the volume of a hemisphere, therefore:

$$\text{volume of a sphere} = \frac{4}{3}\pi r^3$$

(A) Worked examples

a **Volume of desk lamp base**
   A desk lamp will be made with a range of bases. For each of the bases below, calculate the volume of material required. In both cases, the bases are the same height and the central holes for assembly are identical.

   i **Circular base cast in concrete**

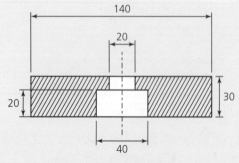

Figure 3.46 Concrete lamp base

The centre line on the drawing of the lamp base cross-section shows that the base is cylindrical. There is a 20 mm diameter hole for the lamp post to attach through and there is a 40 mm diameter hole to fasten the post down with a washer and nut.

Therefore, the calculation involves the subtraction of the volume of the two cylindrical holes from the base cylinder.

volume of a cylinder $= \pi r^2 h$

volume of small hole $= \pi \times 10^2 \times 10 = 3142 \, \text{mm}^3$

volume of large hole $= \pi \times 20^2 \times 20 = 25\,132 \, \text{mm}^3$

volume of the base cylinder $= \pi \times 70^2 \times 30 = 461\,814 \, \text{mm}^3$

volume of concrete $= 461\,814 - 25\,132 - 3142 = 433\,540 \, \text{mm}^3$
$$= 4.3 \times 10^5 \, \text{mm}^3 = 4.3 \times 10^{-4} \, \text{m}^3 \, (2 \, \text{s.f.})$$

Answer: Volume of concrete $= 4.3 \times 10^{-4} \, \text{m}^3 \, (2 \, \text{s.f.})$.

ii **Square base cast in aluminium**

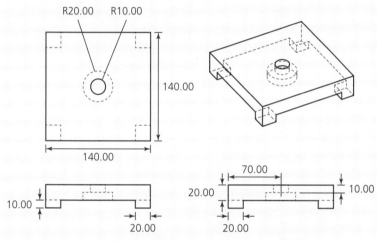

Figure 3.47 Cast aluminium lamp base

The lamp base shown in Figure 3.47 comprises four cuboid feet and the main cuboid. Calculate these separately and add them together to find the total volume without the holes. Then subtract the volume of the holes to obtain the answer.

Step 1: Calculate volume of feet.

volume of one cuboid foot $= 20 \, \text{mm} \times 20 \, \text{mm} \times 10 \, \text{mm} = 4000 \, \text{mm}^3$

volume of four feet $= 4000 \times 4 = 16\,000 \, \text{mm}^3$

Step 2: Calculate volume of main cuboid.

volume of main cuboid $= 140 \, \text{mm} \times 140 \, \text{mm} \times 20 \, \text{mm} = 392\,000 \, \text{mm}^3$

Step 3: Calculate the volume of the holes:

volume of small hole $= \pi \times 10^2 \times 10 = 3142 \, \text{mm}^3$

volume of large hole $= \pi \times 20^2 \times 10 = 12\,566 \, \text{mm}^3$

Step 4: Subtract the volume of the holes from the total of the cuboids to find the volume of aluminium required:

volume of aluminium required $= 392\,000 + 16\,000 - 3142 - 12\,566 = 392\,292 \, \text{mm}^3$
$$= 3.9 \times 10^5 \, \text{mm}^3 \, (2 \, \text{s.f.}) = 3.9 \times 10^{-4} \, \text{m}^3 \, (2 \, \text{s.f.})$$

Answer: $3.9 \times 10^{-4} \, \text{m}^3 \, (2 \, \text{s.f.})$ of aluminium is required to cast the lamp base.

b **Cost of waste material used**

Hi-fi loudspeakers are often fitted with vibration isolation spikes to improve the quality of the acoustics. A manufacturer has designed a new spike that is 25 mm in diameter and 22.5 mm in height (see Figure 3.48). It will be turned from 25 mm diameter stainless steel rod using a centre lathe. The cost of the 25 mm diameter stainless steel rod is £68.95/m. Calculate the cost of the waste material that will be removed by the centre lathe.

The waste material is the difference between the original cylindrical shape of the rod and the cone. A sketch, such as the one in Figure 3.49, generally helps to visualise what is going on.

There are two methods to answer this question. The first is the longest but works through a number of mathematical operations relating to area and volume. The second is the quickest.

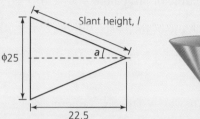

Figure 3.48 Vibration isolation spike

Method 1: Long and uses volume formulae

First, calculate the volume of the waste material.

Calculate the volume of the stainless steel rod used and subtract the spike from it. In other words, calculate the volume of the cylinder and subtract the cone.

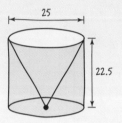

Figure 3.49 Sketch of the problem

$$\text{volume of a cylinder} = \pi r^2 h = \pi \times 12.5^2 \times 22.5 = 11\,044.66 \, \text{mm}^3 \text{ to 2 d.p.}$$

$$\text{volume of a cone} = \frac{1}{3} \text{ base area} \times \text{height} = \frac{1}{3}\pi r^2 h = \frac{1}{3} \text{ volume of cylinder}$$
$$= 11\,044.66/3 = 3681.55 \, \text{mm}^3 \text{ to 2 d.p.}$$

$$\text{volume of waste} = \text{volume of cylinder} - \text{volume of cone} = 11\,044.66 - 3681.55$$
$$= 7363.10 \, \text{mm}^3 \text{ to 2 d.p.}$$

This can be checked:

Remember that a cone is $\frac{1}{3}$ of a cylinder, therefore the waste material must be $\frac{2}{3}$ of the same cylinder.

$$\text{volume of waste} = \frac{2}{3} \text{ volume of cylinder} = \frac{2 \times 11044.66}{3} = 7363.10 \, \text{mm}^3 \text{ to 2 d.p.}$$

Next, calculate the cost per unit volume in units of £/mm³ of the stainless steel rod.

Cost/mm³ = £68.95 ÷ volume of the rod:

$$= \frac{£68.95}{\pi r^2 h} = \frac{68.95}{\pi \times 12.5^2 \times 1000} = £0.000140/\text{mm}^3 \text{ to 3 s.f.}$$

Finally, multiply the cost per volume by the volume of the waste to find the cost of the waste.

$$\text{cost of waste} = 7363.10 \times 0.000140 = £1.03$$

Answer: The cost of the waste stainless steel from turning one spike is £1.03.

Method 2: Quick and relies on use of ratios and memory

If 1000 mm of stainless steel rod costs £68.95, then the cost of 22.5 mm can be calculated as a fraction of this.

$$\text{cost of } 22.5\text{mm} = £68.95 \times \frac{22.5}{1000} = £1.55$$

Remember, the spike is a cone which is $\frac{1}{3}$ of this small section of rod which is a cylinder. This means that the waste is $\frac{2}{3}$ of the small rod and the cost of it must also be $\frac{2}{3}$ of the cost of the small piece of rod.

$$\text{cost of waste} = \frac{2}{3} \times £1.55 = £1.03$$

Answer: The cost of the waste stainless steel from turning one spike is £1.03.

B Guided question

1 **Smart phone packaging**

The packaging for a smart phone will be an acrylic box with a lid and a base. The lid is shown in Figure 3.50. Calculate the volume of acrylic required to injection-mould the lid.

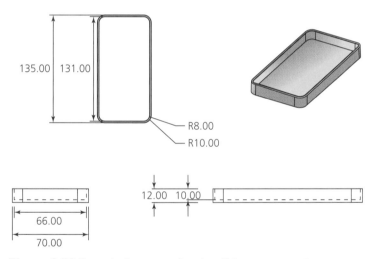

Figure 3.50 Smart phone packaging lid

Break down the calculation into individual three-dimensional parts that are easy to work with – in this case, cuboids and cylinders. Use sketches to plan your workings.

There are different ways of approaching this calculation. These include:

a Consider the lid as a solid object; calculate the volume, then subtract the volume of the middle to make it hollow.

b Calculate volumes of all the components separately, then add them together.

The simplest option is a.

Step 1: Calculate the volume of the solid form.

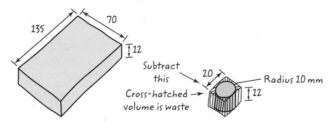

Figure 3.51 Smart phone packaging lid sketched workings

Figure 3.51 shows how you could prepare to do this.

i   Calculate the volume of the lid as a solid cuboid.

ii Calculate the waste lost at each corner to create the outer lid curve.

iii Subtract the waste from the solid.

Step 2: Calculate the volume of the middle.

i   Calculate the volume of the middle as a solid cuboid.

ii Calculate the waste lost at each corner to create the inner lid curve.

iii Subtract the waste from the solid.

Step 3: Calculate the volume of acrylic required

i   Subtract the volume of the middle from the volume of the solid form.

C Practice question

2   **Play area floor protection**

A recent risk assessment of a roundabout at a council-owned park has revealed the hazard of users slipping on the loose sand and soil surface surrounding the children's roundabout. To control the hazard, a high-friction surface that also provides impact resistance is needed. Recycled rubber chippings will be used to provide a low-cost solution.

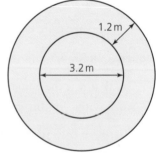

- The roundabout is 3.2 m in diameter.
- The council would like a protective path of rubber chips to surround the ground around the roundabout of width 1.2 m, as shown in Figure 3.52.
- The rubber chips should be laid to a depth of 10 cm.

Calculate the *minimum* volume of rubber chips required. Give your answer in units of m³.

Figure 3.52 Plan view of roundabout with rubber chipping path surrounding it

3.4 Density and mass of three-dimensional objects

It is common for weight calculations to be needed when designing and manufacturing products. It may be necessary to calculate how heavy a product is to determine whether it can be carried, or to ascertain whether a product may fall over if it is top-heavy.

We generally consider how heavy an object is as the weight of the object, and this is how most people refer to weight. In Engineering, however, this is known as the mass. This can be confusing, but it is important to grasp this concept, particularly if you are studying any of the Design Engineering/Engineering Design courses. In Sections 4.4 and 4.5 you will also be working with mass.

TIP

To help you remember:

Mass: measurements of milligrams (mg), grams (g) and kilograms (kg).

- Mass is the amount of matter contained in an object.

- The mass of an object is a constant quantity and does not change with the change of position or location.

Weight: measurements are newtons (N).

- Weight is the force exerted by an object when it is in a gravitational field. It depends upon the gravitational field.

- The weight of an object is the variable quantity and changes with the change in position and location due to the acceleration of the gravity acting on it.

When calculating the mass of an object, the density formula is used:

$$\text{density} = \frac{\text{mass}}{\text{volume}} \qquad \rho = \frac{m}{v}$$

Density is given as mass per unit volume. Common units are $g\,cm^{-3}$ and $kg\,m^{-3}$.

Ⓐ Worked examples

a **Smart phone packaging**

   Look back at the lid of the smart phone packaging on page 53. The density of acrylic is 1.19 g cm^{-3}. Calculate the mass of the acrylic smart phone packaging lid.

$$\text{density} = \frac{\text{mass}}{\text{volume}} \qquad \rho = \frac{m}{v}$$

   Step 1: Convert units into base units of kg, m and m^3.

   The volume can be found from the answer to the question in Section 3.3.

   Volume of acrylic required = 26 459mm³ (nearest mm³) = 2.6459×10^4 mm³ = $2.6459 \times 10^4 \times 10^{-9}$ m³
   $= 2.6459 \times 10^{-5}$ m³

   Density of acrylic = 1.19 g cm⁻³ = 1.19×10^{-3} kg cm⁻³ = $\dfrac{1.19 \times 10^{-3}}{1.0 \times 10^{-6}}$ kg m⁻³ = 1.19×10^{3} kg m⁻³

Step 2: Rearrange the density formula to make mass the subject of the formula.

$$\text{density} = \frac{\text{mass}}{\text{volume}}$$

$$\text{mass} = \text{density} \times \text{volume}$$

Step 3: Calculate the mass of the lid.

$$\text{mass} = 1.19 \times 10^3 \times 2.6459 \times 10^{-5} = 3.15 \times 10^{-2}\,\text{kg (3 s.f.)}$$

Answer: The mass of the lid is $3.15 \times 10^{-2}\,\text{kg}$ (3 s.f.).

b   Mass increase problem

A length of mild steel tubing of 19 mm diameter is used as a handle on a trolley. Product testing has revealed that the tube bends when the trolley is used at its maximum rated loading. The tube has a wall thickness of 1.6 mm. In order to increase rigidity, the manufacturer is considering changing the tube to one with a 3.2 mm wall thickness (see Figure 3.53). The tube is 800 mm long. Calculate the increase in mass of the tube if this change is made.

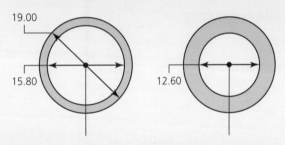

Figure 3.53 Tube cross-section showing increase in wall thickness

For the purpose of this calculation, the density of mild steel is $7700\,\text{kg m}^{-3}$.

To calculate the increase in mass it is necessary to calculate the volume of the tube with 1.6 mm wall thickness (w.t.) and the volume of the tube with 3.2 mm wall thickness.

volume of tube = (area of outer diameter of tube × length) − (area of inner diameter of tube × length)

The area of a circle is: $A = \pi r^2$

volume of 1.6 mm w.t. tube $V = [(\pi \times 9.5^2) - (\pi \times 7.9^2)] \times 800 = 69\,970\,\text{mm}^3$

volume of 3.2 mm w.t. tube $V = [(\pi \times 9.5^2) - (\pi \times 6.3^2)] \times 800 = 127\,071\,\text{mm}^3$

The volume increase is therefore:

$$V_{\text{increase}} = 127\,071 - 69\,970 = 57\,101\,\text{mm}^3$$

To calculate the mass increase, use:

$$\text{mass} = \text{density} \times \text{volume}$$

Convert the density into units of g mm^{-3}.

$$7700\,\text{kg m}^{-3} = \frac{7\,700\,000}{1\,000\,000\,000}\,\text{g mm}^{-3} = 0.0077\,\text{g mm}^{-3}$$

Increase in mass = $0.0077 \times 57\,101 = 440\,\text{g}$

Answer: The mass increase is 440 g.

B Guided questions

Copy out the workings and complete the answers on a separate piece of paper.

1   Batch production costs

The casing of a television remote control is injection-moulded from acrylonitrile butadiene styrene (ABS). 5600 mm³ of ABS is used per casing.

- ■ **The density of ABS is 1.060 g cm⁻³.**
- ■ **The cost of ABS is £0.55/kg.**

Calculate the cost of ABS for 20 000 remote control casings.

Step 1: Convert all values into base units.

Step 2: Calculate the mass of ABS for 20 000 casings.

Step 3: Calculate the cost of ABS for the batch of 20 000.

2   Reading lamp counter-balance

A reading lamp uses a counter-balance that is cast from aluminium with the dimensions shown in Figure 3.54. The density of aluminium is 2720 kg m⁻³. Calculate the mass of aluminium required to cast the counter-balance. Give your answer to the nearest g.

Step 1: To calculate the mass it is first necessary to calculate the volume.

Figure 3.54 shows the side elevation of the counter-balance which for the purpose of the calculation is most easily treated as three separate shapes. Redraw this as shown in Figure 3.55.

The shapes are:

- ■ a hemisphere attached to the base of a truncated cone
- ■ a truncated cone
- ■ a cylindrical blind hole in the narrow end of the truncated cone.

    total volume = volume of hemisphere + volume of cone − volume of cylinder

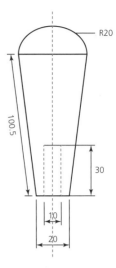

Figure 3.54 Lamp counter-balance

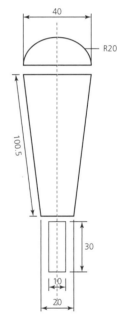

Figure 3.55 Lamp counter-balance calculation

C Practice question

3   Aluminium alloy roof bar

A heavy-duty car roof rack cross bar is tested and found to fail under expected maximum compressive load. The designer has proven that the addition of a second vertical support web, as shown in Figure 3.56 in red, will reinforce the product and solve the problem. The length of the crossbar is 2000 mm.

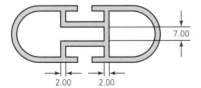 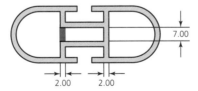

Figure 3.56 Aluminium alloy roof bar

a   Calculate the additional mass created by the support web.
- ■ The density of the aluminium alloy is 270 kg m⁻³.

b   Calculate the additional cost of material required to make the web.
- ■ The cost of aluminium alloy is £1400 tonne⁻¹.

4 Use of trigonometry

Trigonometry is the study of relationships of the lengths of sides and angles of triangles. Triangles are used extensively in the design and manufacture of products and structures.

4.1 Pythagoras

In any right-angled triangle, the length of any side can be calculated if the lengths of the other two sides are known. For the triangle shown in Figure 4.1, Pythagoras' theorem states that:

$$A^2 = B^2 + C^2$$

A is the longest side, known as the hypotenuse.

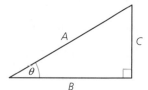

Figure 4.1 A right-angled triangle

This equation can be rearranged to give:

$$A = \sqrt{B^2 + C^2}$$
$$B = \sqrt{A^2 - C^2}$$
$$C = \sqrt{A^2 - B^2}$$

TIP

For a right-angled triangle where the sides are $B = 4$ units in length and $C = 3$ units in length, the hypotenuse A is:

$$A = \sqrt{B^2 + C^2} = \sqrt{4^2 + 3^2} = 5$$

A right-angled triangle with sides in these proportions is often referred to as a '345' triangle.

Note that Pythagoras' theorem is true for right-angled triangles only. In any other triangle, either the sine rule or the cosine rule is required to calculate unknown sides or angles. These rules are covered later in this chapter.

A Worked examples

a Shelf bracket

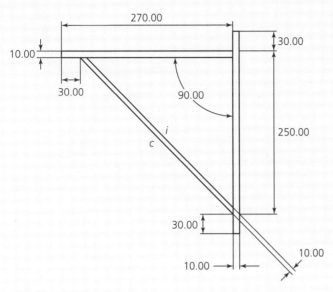

Figure 4.2 Wall bracket

Use the drawing in Figure 4.2 to help you calculate the following dimensions for a wall bracket made of mild steel tubes:

i **Cutting length of diagonal brace, *c*.**

ii **Internal length of diagonal brace, *i*.**

The triangle formed by the bracket is right-angled, so Pythagoras' theorem can be used to calculate the cutting length *c*.

Step 1: The lengths of the horizontal and vertical sides of the triangle are required.

The horizontal side of the triangle = 270 mm − 30 mm = 240 mm long.

The vertical side = 250 mm − 10 mm = 240 mm long.

Step 2: Calculate the length *c* by using Pythagoras' theorem.

$$240^2 + 240^2 = c^2$$

Rearrange to make *c* the subject:

$$c = \sqrt{240^2 + 240^2} = 339.41 \ (5 \text{ s.f.})$$

Answer: The brace should be cut to dimension *c* of 339.41 mm (5 s.f.).

Note: the answer is given to the same accuracy as the dimensions that were provided.

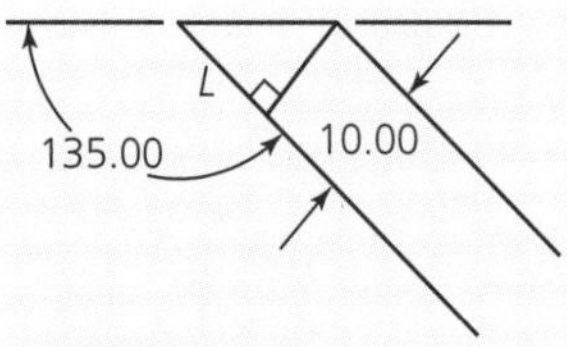

Step 3: Both ends of the brace are cut to the same angle of 45°. Sketch one end of the brace and add the known dimensions to help you. An example is shown in Figure 4.3. In this sketch, *L* is the difference between the internal length, *i*, and the external length, *c*. Therefore:

$$i = c - 2L$$

Figure 4.3 End of diagonal brace

Step 4: Calculate length L.

L could be calculated using trigonometry and you will learn how to do this in Section 4.2. However, you do not need to undertake any calculations if you spot that the triangle is an isosceles triangle with the two acute angles, both of 45°.

This means that L is the same as the other short side of the triangle, i.e. $L = 10.00$ mm.

Step 5: Calculate length i.

$i = c - 2L = 339.41 - 20.00 = 319.41$ mm (5 s.f.)

Answer: The length of i is 319.41 mm (5 s.f.).

b   Calculate the length of the chair legs, A, in Figure 4.4.

The chair legs shown in Figure 4.4 are made of mild steel circular tube. In order to mark out the material for cutting, the distance A is required. The seat base is 550 mm from the floor.

Use Pythagoras' theorem arranged as:

$$A = \sqrt{X^2 + Y^2}$$

Substituting values gives:

$$A = \sqrt{260^2 + 550^2} = 608 \text{ mm to the nearest mm}$$

c   Calculate the separation of the legs, distance B, using Pythagoras' theorem. Assume that there is no gap between the top of the legs beneath the seat base.

The legs of the chair in Figure 4.4 form an isosceles triangle from the floor. This is shown as a red isosceles triangle in Figure 4.5.

The isosceles triangle beneath the chair legs can be treated as two identical right-angled triangles, as shown in orange and green in Figure 4.6. The distance from the seat base to the floor is the same as the hypotenuse, A, of the right-angled triangles that exist between the legs, because it is parallel.

Pythagoras' theorem can be used to calculate the length $\dfrac{B}{2}$ as follows:

$$\frac{B}{2} = \sqrt{A^2 - 550^2}$$

Rearranged to find B, this becomes:

$$B = 2\sqrt{A^2 - 550^2}$$

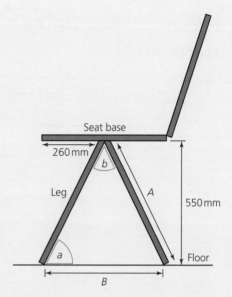

Figure 4.4 Side elevation of chair

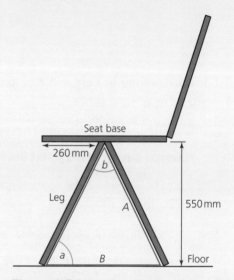

Figure 4.5 Side elevation of chair with isosceles triangle identified

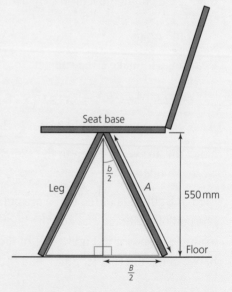

Figure 4.6 Side elevation of chair with right-angled triangles identified

From the answer to part (b) above, the hypotenuse $A = 608\,\text{mm}$, so:

$$B = 2\sqrt{608^2 - 550^2} = 518\,\text{mm}$$

Note that due to the rounding of the value of length A to the nearest mm, this answer is slightly lower than it should be. If the most accurate value of A is used, from your calculator result, the answer is 520 mm.

B Guided question

1 Loudspeaker isolation spike

Look back at Figure 3.48. For completeness of the drawing, slant height, l, is needed. Calculate slant height, l. Give your answer to 1 d.p.

Remember, the isosceles triangle can be divided into two right-angled triangles using the centre line shown in the figure. This enables you to use Pythagoras' theorem.

C Practice question

2 Polypropylene lampshade

The lampshade shown in Figure 4.7 is made from a series of polypropylene strips that run from the top retaining ring to the bottom retaining ring. Gaps in the strips allow light to pass through from the light source.

Figure 4.8 shows the profile of the lampshade. Calculate the length of a single polypropylene strip.

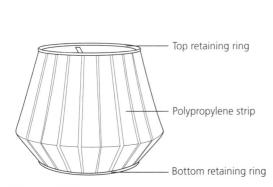

Figure 4.7 Polypropylene lampshade

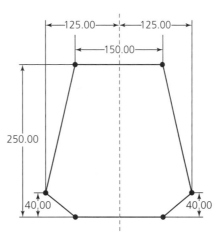

Figure 4.8 Profile drawing of lampshade

4.2 Sine, cosine and tangent

It is important to be able to confidently apply and use the three ratios of the right-angled triangle in order to find the length of a side or size of an angle. These ratios are sine, cosine and tangent (see Figure 4.9).

- The opposite side is always opposite to the angle being calculated.
- The hypotenuse is always the longest side.
- The adjacent is the side next to the angle that is not the hypotenuse.

Useful formulae:

$$\sin\theta = \frac{\text{opposite}}{\text{hypotenuse}}$$

$$\cos\theta = \frac{\text{adjacent}}{\text{hypotenuse}}$$

$$\tan\theta = \frac{\text{opposite}}{\text{adjacent}} = \frac{\sin\theta}{\cos\theta}$$

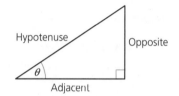

Figure 4.9 Sides of a right-angled triangle for defining sine, cosine and tangent

There are numerous ways to help you remember which formula to use. Here are two:

SOHCAHTOA (it sounds like 'Sockatoa')

or

Some **O**ld **H**ag **C**racked **A**ll **H**er **T**eeth **O**n **A**pples.

Calculating sine, cosine and tangent on your calculator

First, check that your calculator is set to work with degrees rather than radians.

Sin θ can be found on a standard scientific calculator by pressing the sine button and the display will probably show 'sin('. Now enter the angle in degrees followed by ')' and press '='. Your calculator will show the sine of the angle. A similar process will allow you to find the cosine or the tangent.

You will also need to calculate inverse sine, inverse cosine and inverse tangents. On your calculator, these functions look like 'sin^{-1}', 'cos^{-1}' and 'tan^{-1}' respectively. Most calculators require you to press 'shift' to access these functions, or select '2nd' to access the second bank of calculator functions. The display will show 'sin^{-1}'. Now enter the value followed by ')' and press '='. Your calculator will display the value of the angle in degrees. For example, the button presses 'shift', 'sin^{-1}', enter '0.5', followed by button press '=' will display '30', indicating that the angle with a sine of 0.5 is 30°.

Important angles and remembering them

In order to make quick calculations, designers and engineers make use of memorable values for the most popular important angles of 30°, 45° and 60°. The equation of the unit circle in Figure 4.10 can help with this. Pythagoras' theorem gives:

$$x^2 + y^2 = 1^2 = 1$$

Since Figure 4.10 shows that $x = \cos \theta$ and $y = \sin \theta$, this becomes:

$$(\cos(\theta))^2 + (\sin(\theta))^2 = 1$$

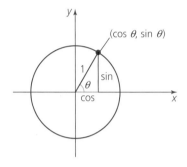

Figure 4.10 Angle calculation

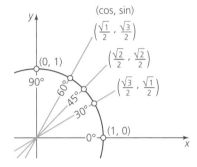

Figure 4.11 Cos and sin values for standard angles

This identity can then be used to compile the table of memorable values shown in Table 4.1.

Table 4.1 Important angles

Angle	sin	cos	$\tan = \dfrac{\sin}{\cos}$
30	$\dfrac{\sqrt{1}}{2} = \dfrac{1}{2}$	$\dfrac{\sqrt{3}}{2}$	$\dfrac{1}{\sqrt{3}} = \dfrac{\sqrt{3}}{3}$
45	$\dfrac{\sqrt{2}}{2}$	$\dfrac{\sqrt{2}}{2}$	1
60	$\dfrac{\sqrt{3}}{2}$	$\dfrac{\sqrt{1}}{2} = \dfrac{1}{2}$	$\sqrt{3}$

TIP

To help you remember:

For sin, think '1,2,3'.

For cos, think '3,2,1'.

A Worked examples

a **Using memorable important angles**

Look back at the wall bracket in the previous section, Figure 4.2. Calculate the length of the diagonal cut end, w, shown in Figure 4.12.

The triangle shown at the end of the brace is an isosceles right-angled triangle and has two sides of 10 mm. All angles are known to be 90°, 45° and 45° respectively. Therefore, the length of w can be calculated by using either the sine or the cosine formula.

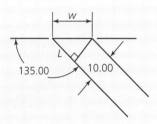

Figure 4.12 Calculating the cut end

$$\sin\theta = \frac{\text{opposite}}{\text{hypotenuse}} \quad \text{or} \quad \cos\theta = \frac{\text{adjacent}}{\text{hypotenuse}}$$

Substituting values gives:

$$\sin 45° = \frac{10\,\text{mm}}{w} \quad \text{or} \quad \cos 45° = \frac{10\,\text{mm}}{w}$$

Remember, both $\sin 45°$ and $\cos 45° = \dfrac{\sqrt{2}}{2}$

Therefore, both formulae become:

$$\frac{\sqrt{2}}{2} = \frac{10\,\text{mm}}{w}$$

Rearrange to make w the subject:

$$w = \frac{2 \times 10\,\text{mm}}{\sqrt{2}} = 14.14\,\text{mm} \;(4\text{ s.f.})$$

Answer: The length of the diagonal cut end, w, of the brace is 14.14 mm (4 s.f.).

b Cutting chair legs

The chair legs shown in Figure 4.4 will be cut from lengths of mild steel circular tube using a slitting saw. The end that is nearest to the floor will be cut at 90°. The opposite end will be welded to the base of the chair. Calculate the angle to which the slitting saw will be set in order for the saw-cut at the top of the leg to be at the correct angle for welding. Give your answer to the nearest degree.

The angle of the slitting saw is the same as angle a because the floor and seat base are parallel to each other. This can be seen with the red overlay shown in Figure 4.13.

There are two ways that this problem could be approached:

If you already had the value of angle b, using the rule of angles in triangles and the symmetry of the chair legs:

$$2a + b = 180°$$

Rearranging would give:

$$a = \frac{180 - b}{2}$$

The other way is to use the sin formula to find angle a:

$$\sin\theta = \frac{\text{opposite}}{\text{hypotenuse}}$$

The opposite is the height of the chair, 550 mm.

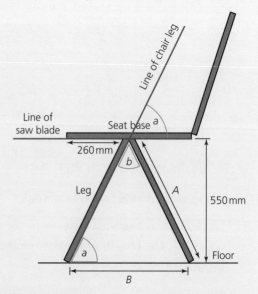

Figure 4.13 Chair leg: slitting saw angle of cut

The hypotenuse, due to symmetry, is the length of leg A, which was calculated to be 608 mm in Section 4.1.

This gives:

$$\sin a = \frac{\text{height}}{\text{length, } A} = \frac{550}{608} = 0.904 \text{ to 3 s.f.}$$

$$a = \sin^{-1} 0.904 = 65°$$

Answer: The angle of the slitting saw will be 65°.

c **Pen pot protective packaging**

Calculate the diameter, d, of the base of the pen pot shown in Figure 4.14. Give your answer to 2 d.p.

The draft angle $a = 5°$.

First, identify that there is a right-angled triangle containing angle a and sketch it as shown in Figure 4.15.

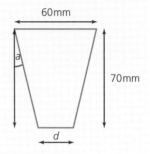

Figure 4.14 Pencil pot

Next, calculate the length of side A, which is the side opposite angle a.

Use the tangent ratio:

$$\tan \theta = \frac{\text{opposite}}{\text{adjacent}}$$

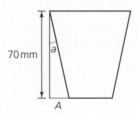

Figure 4.15 Identify the right-angled triangle

Substituting values gives:

$$\tan a = \frac{\text{opposite}}{\text{adjacent}} = \frac{A}{70}$$

In terms of A this becomes:

$$A = 70 \times \tan a = 70 \times \tan 5° = 6.12 \text{ to 2 d.p.}$$

Therefore:

$$d = 60 - (2 \times 6.12) = 47.76 \text{ mm to 2 d.p.}$$

Answer: Diameter, d, = 47.76 mm to 2 d.p.

d **Manufacture of chair leg**

Look back at Figure 4.4. In order to accurately manufacture the chair, the angle between the chair legs, b, is required.

There are two ways that you could approach this problem. If you already had the value of angle a, using the rule of angles in triangles and the symmetry of the chair legs:

$$2a + b = 180°$$

Rearranging gives:

$$b = 180° - 2a$$

From question **b**, above, angle $a = 65°$

$$b = 180° - 130 = 50°$$

Answer: The angle between the chair legs is 50°.

The other way is to use the cos formula to find angle b. Again, consider the isosceles triangle formed by the legs and the floor as two right-angled triangles (see Figure 4.4).

To calculate angle $\dfrac{b}{2}$ the cosine ratio can be used:

$$\cos\theta = \frac{\text{adjacent}}{\text{hypotenuse}}$$

Rearranging gives:

$$\theta = \cos^{-1}\frac{\text{adjacent}}{\text{hypotenuse}}$$

Substituting values gives:

$$\frac{b}{2} = \cos^{-1}\frac{550}{608} = 25°$$

And therefore $b = 50°$.

Answer: The angle between the chair legs is 50°. This agrees with the first answer so it must be correct.

Note that due to the rounding of the value of length a to the nearest mm, this answer is slightly lower than it should be. If the most accurate value of a is used, from your calculator result the answer is 50.6°.

B Guided questions

1 **Manufacture of loudspeaker isolation spike**

   Look back at Figure 3.48. In order to turn the spike on a centre lathe, angle a is required. This is the angle at which the cutting tool and compound slide of the centre lathe will be set for taper turning. Calculate the angle for taper turning, a, to the nearest degree.

   The spike is a cone and to calculate the angle it can be treated as a two-dimensional shape: an isosceles triangle. The centre line creates a right-angled triangle with angle a at the apex. You may want to sketch this triangle before you begin the calculation.

   Decide which of the ratios sine, cosine or tangent is most suitable to solve this problem where the adjacent and opposite are given.

2 **Hexagonal jewellery box**

   The lid of a hexagonal jewellery box is made of solid walnut. The manufacturer believes there is demand for a more decorative version of the lid, and ash veneer

inlays are suggested. See Figure 4.16. Calculate the total cost of ash veneer used for the sides and top of the lid.

- Ash veneer costs £11.10 for a sheet that is 275 cm × 20 cm.

Remember, a hexagon is a regular polygon and made up of triangles.

Calculate the areas of the rectangles and the hexagon, then add them together to find the total area.

3 **Bin geometry**

A recycling bin is injection-moulded from polypropylene. The side profile is shown in Figure 4.17. Calculate the lengths of:

i **line AB**

ii **line CD.**

Step 1: Annotate the drawing to identify what you know and what you need to find out.

Step 2: In order to calculate the length of line AB, the length of line y is needed. To calculate this, the length of line x is needed (x and y from your annotated drawing from Step 1).

Step 3: Calculate line y.

Step 4: Calculate the length of line AB.

Step 5: Calculate the length of line CD.

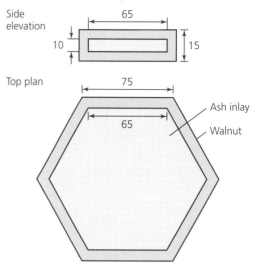

Figure 4.16 Hexagonal jewellery box

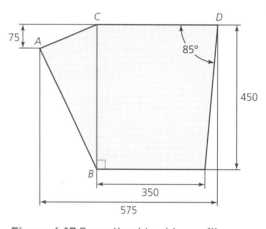

Figure 4.17 Recycling bin side profile

ⓒ Practice questions

4 Manufacture of desktop

Look back at the parallelogram-shaped desktop question at the end of Section 3.1 (page 41). 19 mm oak-veneered MDF will be used for this desktop, at a cost of £69 for a standard-sized board that is 1220 mm × 2440 mm (4′ × 8′).

a Calculate the minimum size of rectangle from which the desktop could be cut. Give your answer to the nearest mm.

b Calculate the cost of the desktop to the nearest penny.

The desktop will be protected with lacquer on the top face and all vertical faces. 1 litre of lacquer costs £32 and covers a minimum of 8 m².

c Calculate the cost of lacquer required to provide two coats to the desktop.

5 Dining table legs

A manufacturer has had a new dining table designed. It will have a 10 mm toughened-glass top and solid oak legs. The designer has used anthropometric data to calculate the height and depth of the table, as shown in Figure 4.18. The length of the table legs now needs to be calculated.

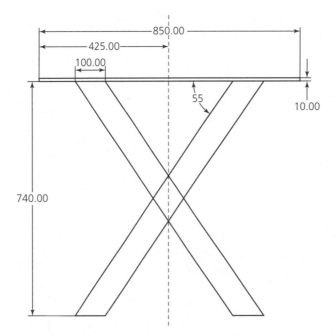

Figure 4.18 Dining table end elevation dimensions

Calculate the length of oak required for each table leg before the ends are cut to the 55° angle.

6 Volume of star-shaped lamp base cast in aluminium

A desk lamp will be made with a range of bases. One option is a star-shaped base cast in aluminium, as shown in Figure 4.19. The lamp base will be 30 mm in height.

In Section 3.3, worked example **ai** on page 50, the volumes of the mounting holes were calculated to be:

volume of small hole = 3142 mm³

volume of large hole = 25 132 mm³

Calculate the volume of aluminium required to cast the base.

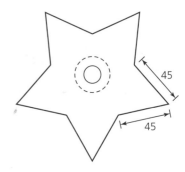

Figure 4.19 Star-shaped aluminium base

4.3 Sine and cosine rules

The sine and cosine rules are often required in order to find the length of a side or size of an angle of a triangle that is not right-angled. For right-angled triangles, Pythagoras' theorem can be used. This is covered in Section 4.1.

The sine rule

For a triangle with sides a, b and c and angles A, B and C, as shown in Figure 4.20, the sine rule is:

$$\frac{a}{\sin A} = \frac{b}{\sin B} = \frac{c}{\sin C}$$

or

$$\frac{\sin A}{a} = \frac{\sin B}{b} = \frac{\sin C}{c}$$

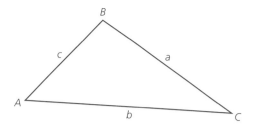

Figure 4.20 Non-right-angled triangle labelled for use with the sine and cosine rules

Use it when the following are given:

- two sides and an angle opposite to one of the two sides (see Figure 4.21).
- two angles and a side opposite one of the angles (see Figure 4.22).

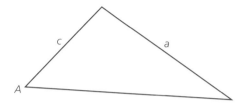

Figure 4.21 Two sides and one opposite angle are given

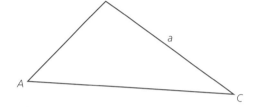

Figure 4.22 One side and two angles are given

The cosine rule

For the triangle shown in Figure 4.20 the cosine rule is:

$$a^2 = b^2 + c^2 - 2bc \cos A \quad \text{or} \quad b^2 = a^2 + c^2 - 2ac \cos B$$

or $c^2 = a^2 + b^2 - 2ab \cos C$

These formulae can be rearranged to give:

$$\cos A = \frac{b^2 + c^2 - a^2}{2bc}$$

$$\cos B = \frac{a^2 + c^2 - b^2}{2ac}$$

$$\cos C = \frac{a^2 + b^2 - c^2}{2ab}$$

Use it when the following are given:

- two sides and an angle adjacent to a side (see Figure 4.23a).
- three sides (see Figure 4.23b).

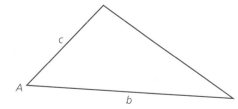

Figure 4.23a Two sides and an adjacent angle are given

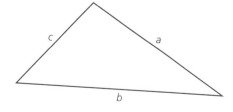

Figure 4.23b Three sides are given

Ⓐ Worked examples

a **Occasional table**

Calculate the lengths of the edges of the legs labelled *A* and *B* in Figure 4.24.

Step 1: A triangle exists beneath the chair legs with sides *A*, 70 mm and 370 mm.

The angle between the 70 mm and 370 mm sides of the triangle can be found by subtracting (7° + 5°) from 90° = 78° (see Figure 4.25). You should annotate the original drawing or sketch this triangle to help your calculations.

Step 2: Use the sine rule to calculate *A*.

$$\frac{A}{\sin 78°} = \frac{370}{\sin 91°}$$

Rearrange to make *A* the subject:

$$A = \frac{370}{\sin 91°}\sin 78° = 362.0\text{mm}\ (1\text{ d.p.}) = 362\text{mm}\ (3\text{ s.f.})$$

Answer: *A* is 362 mm long (3 s.f.).

Step 3: The cosine ratio can be used to calculate the length of *B*.

$$\cos\theta = \frac{\text{adjacent}}{\text{hypotenuse}}$$

Substitute values:

$$\cos 7° = \frac{610\text{mm}}{(B + 40)}$$

Rearrange to make *B* the subject:

$$B = \frac{610}{\cos 7°} - 40 = 574.58 = 575\text{mm (3 s.f.)}$$

Answer: B is 575 mm long (3 s.f.).

b **Geometric logo design**

Figure 4.26 shows a geometric design that will be embroidered on a garment as a company logo. In order to accurately mark out the logo, angle *C* is required. Calculate angle *C*.

Use the cosine rule:

$$\cos C = \frac{a^2 + b^2 - c^2}{2ab}$$

The left 'leg' of the logo is a mirror image of the right 'leg', so:

$$a = 34\text{mm} \quad b = 29\text{mm} \quad c = 20\text{mm}$$

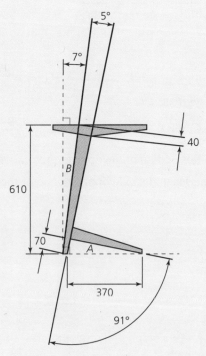

Figure 4.24 Side elevation of occasional table: *A* and *B* are distances of edges on legs

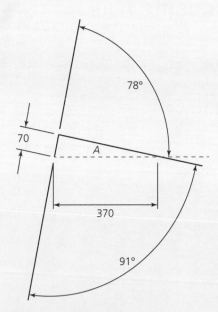

Figure 4.25 Calculating *A*

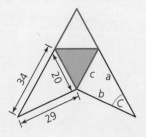

Figure 4.26 Embroidered logo

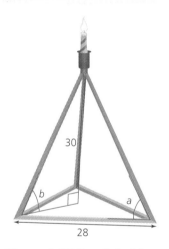

Substituting values gives:

$$\cos C = \frac{34^2 + 29^2 - 20^2}{2 \times 34 \times 29} = 0.81$$

On your scientific calculator, use the inverse cosine button, probably identified as 'cos⁻¹' or 'acos', to calculate C.

$$C = \cos^{-1} 0.81 = 35.9°$$

Answer: Angle $C = 35.9°$.

c   **Look back at Figure 4.4. Calculate angle b using the cosine rule.**

Note that in this example the angle is b and lengths are A, B and C. Length C is the same as A. The lengths of A and B were calculated in previous worked examples and are:

$$A = 608\,\text{mm and } B = 520\,\text{mm}$$
$$\cos b = \frac{A^2 + C^2 - B^2}{2AC} = \frac{608^2 + 608^2 - 520^2}{2 \times 608 \times 608} = 0.63 \text{ to 3 s.f.}$$
$$b = \cos^{-1} 0.63 = 50.6°$$

Answer: Angle $b = 50.6°$ to 3 s.f.

B Guided questions

1   **Candleholder geometry**

A manufacturer has a new design of candleholder with a pyramid frame stand made of square section mild steel rod. Figure 4.27 shows the prototype. In order to manufacture a batch of 10 000 of these candleholders, jigs need to be made and the angles a and b are required. Calculate angles a and b.

In order to calculate these angles, the length of the diagonal uprights is required. This can be calculated by using the vertical right-angled triangle that has the 30 mm vertical dimension and angle b, shown in Figure 4.28.

Figure 4.27 Candleholder

The line labelled as Centre, C, can be calculated using your knowledge of the properties of the base shape, which is an equilateral triangle. A sketch like the one in Figure 4.29 will help you.

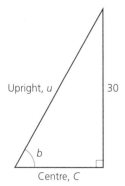

Figure 4.28 Vertical triangle

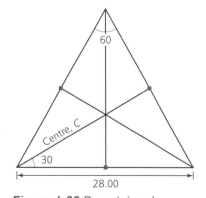

Figure 4.29 Base triangle

Once the length of line C is calculated, you should be able to calculate the length of the diagonal upright, u. From this you will be able to calculate the angles a and b.

To help you calculate angle a, a drawing of the right-angled triangle that exists within the side isosceles triangle, like the one in Figure 4.30, will help you.

Step 1: Calculate centre, c.

The cosine ratio can be used to calculate this.

Step 2: Calculate u.

Pythagoras' theorem can be used to calculate u.

Step 3: Calculate angle a using the cosine ratio again.

Step 4: You now know all of the sides of the triangle containing angle b, so the cosine rule can be used to calculate b.

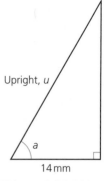

Figure 4.30 Side triangle

2   **Laser cutting acrylic**

A lightning strike shape will be laser cut from white 3 mm acrylic. The laser cutter moves at a speed of 3.5 mm s⁻¹ when cutting 3 mm acrylic. Using Figure 4.31 with dimensions to help you, calculate the time it will take to cut one lightning strike out of 3 mm acrylic.

Step 1: Annotate the drawing of the lightning strike to identify the dimensions that you need to find using trigonometry. Figure 4.32 shows how the lightning strike can be annotated..

Step 2: The shape comprises a trapezium and a triangle. The angle 110° can be identified because the rules of a triangle mean that all angles should add up to 180°, i.e. $50 + 20 + 110 = 180$. The triangle is not right-angled, but two angles and one side are known. Use the sine rule to calculate b.

Step 3: Use the sine rule to calculate the combined dimension of $a + c$.

Step 4: It is now necessary to divide the quadrilateral into triangles. By drawing in a diagonal as shown in Figure 4.33, the resulting hypotenuse can be calculated using Pythagoras' theorem.

Step 5: Use Pythagoras' theorem to calculate the length of d.

Step 6: Use the sine rule to calculate angle e. This will help you calculate angle f, which is needed to calculate the length of the side of the triangle that is $(50\,\text{mm} + a)$.

Step 7: Calculate angle f using your knowledge of the rule that all internal angles of a triangle add up to 180°.

Step 8: Use the sine rule again to calculate the length of the side $(a + 50)$.

Step 9: Calculate the length of line a.

Step 10: Calculate the length of line c by subtracting a from $(a + c)$ calculated in Step 3.

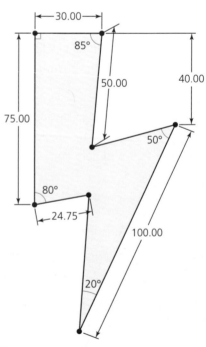

Figure 4.31 Lightning strike with dimensions

Step 11: Calculate the total length of the perimeter of the lightning strike.

Step 12: Calculate the time taken to cut the perimeter.

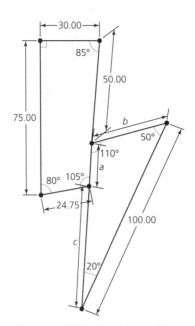

Figure 4.32 Lightning strike annotated first stage

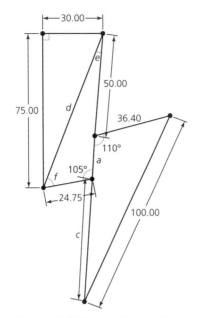

Figure 4.33 Lightning strike annotated second stage

C Practice question

3 Tipi ground sheet

A six-person tipi, shown in Figure 4.34, has been designed to have a regular decagonal seam welded polyester groundsheet, as shown in Figure 4.35. Calculate the area of the groundsheet. Measurements are in centimetres.

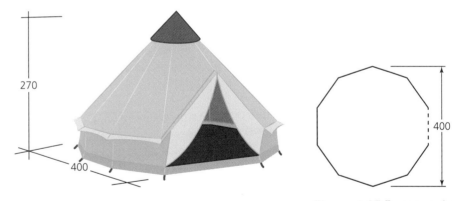

Figure 4.34 Tipi

Figure 4.35 Decagonal groundsheet

Sections 4.4 Direction of movement and 4.5 Resolving force vectors are available online at www.hoddereducation.co.uk/essentialmathsanswers

5 Use and analysis of data, charts and graphs

In this chapter you will be introduced to some simple principles to help you present information and data clearly. You will learn how to improve readers' understanding of the data in order to best convey the messages held within it. In Chapter 7 you will learn how to analyse and interpret data.

5.1 Presenting data

Data are a collection of facts in numerical form. These can be values or measurements and within the collection lies information that is waiting to be shared. This is done through interpretation, data presentation and statistics. It is essential that data are fully understood before they can be used effectively.

Types of data

The two main types of data are **qualitative** and **quantitative**.

Qualitative data (descriptions, opinions etc.)

Qualitative data are observed but not measured. They are often subjective and open to the interpretation of an individual's bias. These data are normally descriptive and the sort of response you would gain from stakeholder interviews — for example: the trolley was hard to manoeuvre. Here the response is subject to the size, strength and experience of the interviewee.

Another example of qualitative data would be the names of all the flavours of ice cream sold by a vendor. Quantitative data would be the number of ice creams sold in each flavour.

The types of qualitative data that you are most likely to encounter are:

- Categorical or nominal — this refers to named categories of data, e.g. wall-mounted or floor standing. There is no natural order to these data.

- Ordinal — e.g. small, medium, large. These data have natural order.

Quantitative data (numbers or values)

Quantitative data are numerical and deal with commodities that are measurable, such as load, force, height, width, length, area, volume, temperature etc. — for example, the desk is 750 mm high × 900 mm wide × 2000 mm long. A range of tables could be compared using a comparison of the values of these dimensions.

The types of quantitative or numerical data that you are most likely to encounter are:

- integers — countable whole numbers, e.g. 1, 2, 3… — these are also real numbers

- real or rational numbers — these are all the positive, negative, fraction and decimal numbers, e.g. 0.34, −0.5, $\frac{1}{3}, -\frac{7}{8}$

- irrational numbers — these are usually decimals that do not terminate or repeat. They go on for ever, e.g. $\sqrt{2}$ and π.

Presenting numbers in text

There are four general guidelines for presenting numbers in text. These are intended to help the reader and convey your message with minimal disruption and optimum effectiveness in the interpretation of the numbers. There is a common misconception that the most accurate number is best, but in fact this can lead to readers questioning the accuracy of the data, particularly if very large numbers are being quoted to include single units.

For example, if the results of a large survey of user opinions indicated that 7523 users were unhappy with the performance of their product, the data presented should be rounded down to 7500. It is unrealistic to assume that a survey of this magnitude could be accurate to the nearest person.

You should aim to follow these examples of good practice in your NEA.

1 Numbers from zero to nine should be written in words rather than digits, e.g. 'four'.

2 Numbers larger than nine should be written using digits, e.g. '14' and '51'.

3 Numbers up to 10 000 should be written rounded to the nearest hundred, e.g. '7452' should be presented as '7500'.

4 Numbers over 10 000 should be presented using a mixture of digits and words. For example, 12 564 should be written as 13 thousand.

C Practice question

1 Present the following numbers in a form that is suitable for inclusion in text:

a 5

b 99 887

c 150 033

d 2

e 23

5.2 Statistics

Statistics is an area of mathematics that involves:

- collecting data

- presenting data

- analysing data

- interpreting data.

Statistics deals with the collection, analysis and interpretation of data. In this section you will learn how to use statistics to present useful information that could help a designer or manufacturer make decisions.

In order to present information effectively in a simple form, **summary statistics** are used to summarise the observations made of sets of data.

The most common statistics that you will encounter are measures of:

- location

- spread

- shape.

Location

Measures of location in the data set include mean, median and mode. You may need to calculate or present these measures of location from lists of numbers or from the analysis of graphs. It is important that you know the differences between these three measures and how to use them.

It is possible for the mean, median and mode to be the same value for a particular data set, but this is more often not the case. To learn why this is so, look at the '**shape of graphs**' section later in this chapter.

Mean

This is what is commonly known as the 'average'. To calculate the mean, add up all the values and divide by the number of values.

 Worked example

The heights, in mm, of a sample of desks measured were 740, 777, 745, 755, 750, 750, 742, 780. Calculate the average height.

For this sample:

$$\text{mean average} = \frac{\text{sum of values in sample}}{\text{number of values in sample}}$$

$$\text{mean average} = \frac{740 + 777 + 745 + 755 + 750 + 750 + 742 + 780}{8}$$

$$= 754.875$$

Answer: The average (mean) height of the desks is 755 mm (3 s.f.).

Median

This is the value that falls in the middle of the sample. To find the median, numbers in a list need to be arranged in numerical order from smallest to highest.

 Worked example

Find the median value of the desk heights in the worked example above.

The sample becomes:

740, 742, 745, **750, 750**, 755, 777, 780

The middle values are both 750, so the median is 750.

For a sample with an odd number of values, the median is the middle value. For a sample with an even number of values, the median is the mean of the middle two values.

Ⓐ Worked examples

Calculate the median for the following samples of measurements:

a   23, 24, **25**, 26, 27

(number of values + 1) ÷ 2 = (5 + 1) ÷ 2 = 3

The middle number is the third, so **25** is the median.

Answer: 25 is the median.

b   10, 11, 12, **13**, **14**, 15, 16, 17

(number of values + 1) ÷ 2 = (8 + 1) ÷ 2 = 4.5 or between the 4th and 5th value

So, the average of the fourth and fifth values of **13** and **14** is required. This is 13 + 14 = 27 ÷ 2 = 13.5.

Answer: 13.5 is the median.

Mode

This is the value that occurs the most often. If there is no repetition in the data sample, there is no mode.

Ⓐ Worked examples

a   **What is the mode of the original sample of table heights?**

740, 742, 745, **750**, **750**, 755, 777, 780

Answer: 750 mm occurs twice, so 750 is the mode.

b   **For the following sample of measurements, what is the mode?**

10, 11, 12, 13, 14, 15, 16, 17

Answer: There is no repetition, so there is no mode.

Ⓑ Guided question

1   **A survey of laptops weighing less than 1 kg has been undertaken and the results have been collated in Table 5.1. Identify the modes for:**

a   **type**

b   **screen size**

c   **CPU.**

Table 5.1 Laptop survey results

Model	Type	Screen	CPU	Weight (kg)
Samsung Galaxy Tab Pro S	Tablet	12.1″	Intel Skylake Core M	0.69
Microsoft Surface Book	Tablet	13.5″	Intel Skylake Core U	0.73
Lenovo Miix 700	Tablet	12.0″	Intel Skylake Core M	0.77
Microsoft Surface Pro 4	Tablet	12.3″	Intel Skylake Core U	0.78
Microsoft Surface Pro 3	Tablet	12.0″	Intel Haswell Core U	0.79
Samsung Notebook 9 900X3L	Clamshell	13.3″	Intel Skylake Core U	0.84
Lenovo Lavie Z	Clamshell	13.3″	Intel Broadwell Core U	0.84
Sony Vaio Pro 11	Clamshell	11.6″	Intel Haswell Core U	0.87
Apple Macbook 12	Clamshell	12.0″	Intel Skylake Core M	0.91
Asus Chromebook C201	Chromebook	11.6″	Rockchip	0.91
Asus Chromebook Flip	Chromebook	10.1″	Rockchip	0.91
Asus Zenbook UX390UA	Clamshell	12.5″	Intel Kabylake Core U	0.91
Lenovo Lavie Z 360	Convertible	13.3″	Intel Broadwell Core U	0.92
Samsung ATIV Book 9	Clamshell	12.0″	Intel Core M	0.95
LG Gram 13	Clamshell	13.3″	Intel Skylake Core U	0.98
LG Gram 14	Clamshell	14.0″	Intel Skylake Core U	0.98
LG Gram 15	Clamshell	15.6″	Intel Skylake Core U	0.98
Lenovo Yoga 900S	Convertible	12.5″	Intel Skylake Core M	0.99
HP EliteBook Folio G1	Clamshell	12.5″	Intel Skylake Core M	0.99

The mode is the value with the most occurrences. It is also known as the most popular value in the sample of data.

Ⓒ Practice question

2 For the laptop weights in Table 5.1:

    a Calculate the mean (average) of the laptops surveyed.
    b Identify the laptops that best represent the median of the sample.
    c Identify the mode of the sample.

Mode in charts and graphs

The sample of desk measurements is represented in Figure 5.1 in a line graph. The **frequency** represents how many times each value of measurement of the height is present in the sample of data. For example, 740 mm appears only once, hence it has a frequency of 1 on the graph.

The benefit of the graph is that it is easy to identify the mode. It is clear to see that the mode is the value that corresponds with the highest point of the graph.

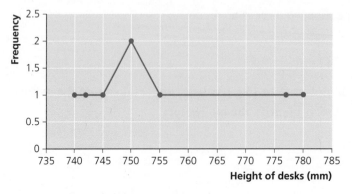

Figure 5.1 Desk height survey results

Full worked solutions at **www.hoddereducation.co.uk/essentialmathsanswers**

In fact, on any graph it is easy to identify or find the mode, from the highest point, or joint highest points, on the graph.

Ⓐ Worked example

The bar graph in Figure 5.2 represents data gathered from a larger sample. What is the most popular height of desk?

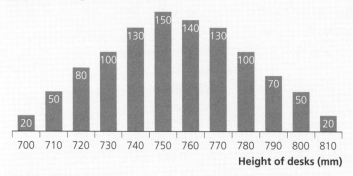

Figure 5.2 Desk height survey results — bar graph

Answer: The graph has an easily identifiable peak in frequency at 750 mm. Therefore, the mode is 750 mm.

Comparison of location on a graph

Graphs and charts can be used to compare the distribution of two sets of data. For example, the two distributions shown in Figure 5.3 are identical in spread and shape, but their location on the *x*-axis is different. The value in red is always more likely to be higher than the value in green. This could be a comparison of anthropometric data, such as hand span, for girls and boys at the age of 14, where the values for the boys, in red, are in general larger than the values of the girls, in green.

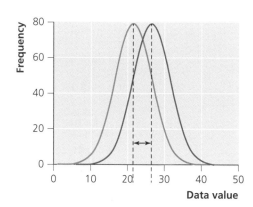

Figure 5.3 Comparing location

Spread

On a graph, spread is useful to show a comparison of range between two sets of data, as shown by the arrows in Figure 5.4. The greater spread of the red group means that the values have a greater variation than those of the green group.

In this example, the data could represent the sales of two products. The product represented by the green line gains popularity quickly, then sales drop off quickly. The sales of a seasonal product would look like this — for example, Halloween outfits. The product represented by the red line sells for a longer period of time — for example, the sales of festival tents over the summer months.

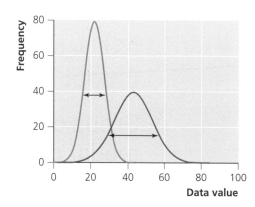

Figure 5.4 Comparing spread

Shape of graphs

The shape of a graph can tell a story far quicker than the equivalent data in a table, for example the graph in Figure 5.5 shows the comparison of the age at failure and disposal of two models of toasters, over time. From this chart, it can easily be seen that the red toaster has outlasted the green toaster, with failures spread out over a much greater length of time.

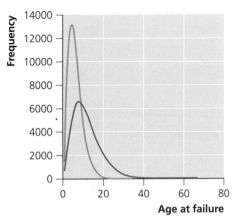

Figure 5.5 Comparing shape

Skewness and the effect of shape on mean, median and mode

Skewness is the way that the data set is distributed to the left or right of the graph. The data represented in Figure 5.5, for example, skew towards the left. This is known as **positive skew**.

Skew has an effect on the mean, median and mode, as shown in Figure 5.6. It is good to be familiar with the relationship between mean, median and mode for different skews. This can help you easily check your calculations.

The **symmetrical distribution** is a classic example of a bell-shaped distribution, typical of the shape of a graph representing any anthropometric data, such as hand span. Due to the symmetry of the distribution, the mean, median and mode are all the same.

For a data set with **positive skew**: mode < median < mean.

For a data set with **negative skew**: mean < median < mode.

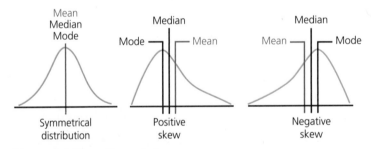

Figure 5.6 The effect of skew

5.3 Group data, estimates, modal class and histograms

Group data, estimates and modal class

Sometimes it is necessary to group data when there is a lot of it. A good example of this is age. For example, Table 5.2 groups the ages of a sample of consumers of luxury electric fans. The groups are known as classes.

Table 5.2 Frequency: electric fan users

Age	Frequency
80–89	50
70–79	300
60–69	740
50–59	1150
40–49	1200
30–39	750
20–29	430
10–19	20

The class **upper limits** are the highest values in each class, e.g. 19, 29, 39, 49, 59, 69, 79 and 89.

The class **lower limits** are the lowest values in each class, e.g. 10, 20, 30, 40, 50, 60, 70, 80.

When grouped data are presented in tables and graphs it is a little trickier to calculate the mean, which can only be an **estimate**. The calculation of the median and identification of the mode are similar to the methods used when working with standard frequency tables and graphs.

The class with the greatest frequency is called the **modal class**.

 Worked examples

a **Calculate the mean age of the electric fan users in Table 5.2.**

Consider the class 80–89:
- 50 users are aged 80, 81, 82, 83, 84, 85, 86, 87, 88 or 89, but it is not known *exactly* how old each user is.
- If the age of 89 is assumed, it is likely that the *estimate* of the mean age would be too high.
- If the age of 80 is assumed, it is likely that the *estimate* of the mean age would be too low.
- It is therefore better to use the mid-point of the class and assume that the age of each user is 84.5. By using this *estimate*, the age is more likely to reflect the mean.

The mid-points of the other classes should be found as shown in Table 5.3.

Table 5.3 Calculating the mean age of users

Age	Frequency, f	Mid-point age, x	fx
80–89	50	84.5	4 225
70–79	300	74.5	22 350
60–69	740	64.5	47 730
50–59	1 150	54.5	62 675
40–49	1 200	44.5	53 400
30–39	750	34.5	25 875
20–29	430	24.5	10 535
10–19	20	14.5	290
Total:	4 640		227 080

For this set of data:

$$\text{Mean average} = \frac{\text{sum of values in sample}}{\text{number of values in sample}} = \frac{\text{sum of } fx}{\text{sum of } f} = \frac{227080}{4640} = 48.94 = 49 \ (2 \text{ s.f.})$$

This is an estimate.
Answer: The estimated mean age of users in the sample is 49.

b **Calculate the median age of the electric fan users sampled in Table 5.2.**

The median is the middle value when the values in a set of data are arranged in order of size. The data are grouped into classes, so it is not possible to find an exact value for the median. It is possible to identify the class that contains the median.

For this set of data:

$$\text{median value} = (\text{number of values} + 1) \div 2 = (4640 + 1) \div 2 = 2320.5$$

Therefore, the class containing the 2320th or 2321st value contains the median.

With a large set of data like this you may need to add up entries in the frequency column to be sure of identifying the correct class.

$20 + 430 + 750 = 1200$, which is smaller than 2320. The next frequency entry for the 40–49 age group is 1200 and the sum becomes 2400.

Therefore, the class containing the median must be the 40–49 age group.

Answer: The median age is contained within the 40–49 age class.

c **What is the modal class in Table 5.1?**

The class with the most people is aged 40–49.

Answer: The modal class is age group 40–49.

Histograms

Histograms are similar to bar graphs, but they are used to represent values that are grouped into ranges, or as mentioned above, **classes**. This allows a set of continuous data to be presented so that the distribution can be analysed.

In a bar chart, frequency is represented by the vertical height of the bar and the y-axis of the graph should be labelled as frequency. It is different for a histogram. For a histogram, it is the *area* of the bar that represents frequency and the y-axis label should be **frequency density**.

 ## Worked example

Look back at Table 5.1, which provides the results of a survey of laptops weighing less than 1 kg. Present the weight data in a histogram.

Step 1: To draw the histogram, first identify the **class boundaries**. Suitable lower boundaries for weight would be of 0.6 kg, 0.7 kg, 0.8 kg and 0.9 kg.

Step 2: Calculate the **class widths**.

In this case all the class widths are the same, i.e. $0.7 - 0.6 = 0.1$ is the same as $0.9 - 0.8 = 0.1$. The class widths are 0.1 kg.

Step 3: Calculate the **frequency density**.

$$\text{area} = \text{frequency} = \text{frequency density} \times \text{class width}$$

Therefore:

$$\text{frequency density} = \frac{\text{frequency}}{\text{class width}}$$

It is a good idea to create a table when you calculate these densities, similar to the one in Table 5.4.

Table 5.4 Data for histogram

Weight (kg)	Lower class limit	Frequency	Frequency density
0.9–0.99	0.9	11	110
0.8–0.89	0.8	3	30
0.7–0.79	0.7	4	40
0.6–0.69	0.6	1	10

Step 4: Draw the histogram, as in Figure 5.7.

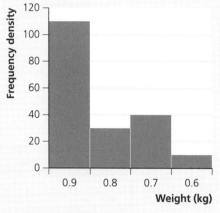

Figure 5.7 Histogram

B Guided question

1   **Table 5.3 contains data that has been grouped from a continuous set of data, in this case, the age of users sampled as part of market research. Present the data in Table 5.3 in a histogram.**

Step 1: To draw the histogram, first identify class boundaries.

Step 2: Calculate the class widths.

Step 3: Calculate the frequency density.

Step 4: Draw the histogram.

5.4 Presenting market and user data

TIP

Histograms can be used to represent data with unequal class widths as shown in Figure 5.8. These work in a similar way.

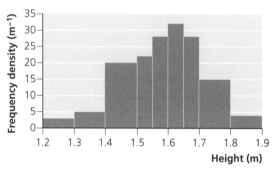

Figure 5.8 A frequency–density histogram of the heights of a sample of people

Questionnaires and surveys are commonly used to gather important information that will help identify stakeholder requirements and steer the creative phase of any project. When you undertake research for your NEA, if you gather a large quantity of information, or data, it is generally best to **collate** it into a visual form in a chart or a table.

TIP

The word **collate** means to collect and combine, and usually refers to texts, information and data.

The benefit of tables and charts is that they can present information in a format that is easy for the reader to understand. You will be familiar with the old adage 'a picture is worth a thousand words'. This can also be the case with presentation of data. However, often data are presented in a confusing form that misleads the reader.

Example 1 presents data from a table in a graphical form that helps the reader, while Example 2 serves to demonstrate how data can be presented in a confusing way.

Example 1

A manufacturer would like to determine consumers' preferred surface finish for a new laptop-type product. The surface finishes being considered are black, silver, space grey and white.

If you asked a sample of 40 representative consumers of the laptop which surface finish they would prefer, you would probably use a spreadsheet application to **collate** your results into a table, as in Table 5.5.

Table 5.5 Survey results

Surface finish	Frequency
Black	2
Silver	12
Space grey	21
White	5
Total	40

TIP

Most spreadsheet applications will enable you to use summing tools to automatically add up the entries in each column. You can also write your own formulae to undertake automatic calculations such as averages and percentages.

Most applications have chart wizards that allow you to quickly and easily present the same data in a chart, such as the bar chart in Figure 5.9. But beware! As the examples here illustrate, the ease with which data can be presented by spreadsheet applications can lead to bad judgement.

Bar graphs

The bar graph in Figure 5.9 is a good example of how the visual presentation of the data from a table has been improved to enable the reader to instantly identify the most popular choice of space grey. In addition, the values above each column save the reader from having to deduce the value of each entry on the y-axis. Differences in values are also easy to read and the data can be analysed and interpreted further.

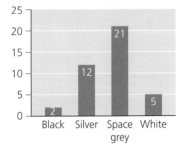

Figure 5.9 Survey results: a good example

Example 2

The three-dimensional bar chart in Figure 5.10 has been generated by the same spreadsheet application using the same data as that used to generate Figure 5.9. However, although the reader can easily see that space grey is the most popular surface finish, it is difficult and misleading for them to interpret what the corresponding value of frequency should be from the y-axis. Figure 5.11 is the same chart, edited to identify the possible values that the reader could consider to be valid for space grey.

If the reader looks across to the y-axis from the back edge of the three-dimensional bar representing space grey, they will probably follow the path of the red line. This would seem to indicate a value between 17 and 20 on the y-axis. Alternatively, reading from the leading edge of the three-dimensional bar would result in the

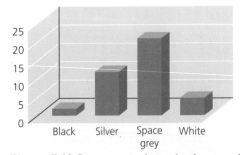

Figure 5.10 Survey results: a bad example

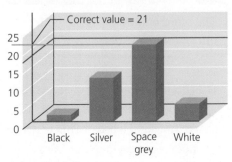

Figure 5.11 Survey results: a bad example annotated

green path, which indicates a value between 23 and 25. In both cases the interpretation of the data results in an incorrect value, with inaccuracy of as much as 10%. The correct value can be extracted from this graph only by following the front edges of the three-dimensional bars across to the *y*-axis and intersecting the green line, as shown by the blue line. This is extremely confusing and it is possible that you still have not figured this out, even after explanation.

This is a perfect example of how you could be tempted to get carried away and present the data in what you perceive to be an exciting visual form. Unfortunately, in this case, as in most cases when presenting data, *simpler is definitely clearer.*

In the following sections you will learn some ground rules to help you create tables, charts and graphs that make data both interesting and appropriate for analysis and interpretation.

Tables

There are a couple of golden rules when using tables to present data to a reader.
Use tables when:

- identification of trends is not important
- the number of values is small.

Presenting numbers

It goes without saying that *consistent formatting* in a table is important to avoid confusion.

- Keep the data in columns in line with the same formatting as shown in Tables 5.6 and 5.7, which present typical sales data.
- Use consistent scales and units, as well as a consistent number of decimal places or significant figures.

Table 5.6 presents a number of years' sales to a high degree of accuracy. This is not necessary to get the message across and, in fact, makes the table quite hard to interpret. By presenting the sales to a lower degree of accuracy, in millions, to one decimal place, as in Table 5.7, a trend is instantly identifiable. The trend is that sales have been increasing since 2012.

Design rules: keeping it simple

1 Sorting:
   a A table should read from left to right.
   b Time should count from past to present, as shown in Table 5.6.
2 When comparing numbers, it is easier to read the numbers in columns, as in Table 5.6.
3 Totals:
   a Row totals should be at the right-hand side.
   b Column totals should be at the bottom, as shown in Table 5.5.

Table 5.6 Year-end sales

Year	Sales (units)
2012	2 103 432
2013	2 457 650
2014	2 899 765
2015	3 267 589
2016	3 567 882
2017	3 876 545
2018	4 067 546

Table 5.7 Year-end sales

Year	Sales (million units)
2012	2.1
2013	2.5
2014	2.9
2015	3.3
2016	3.6
2017	3.9
2018	4.1

More charts and graphs

Charts and graphs are a better alternative to tables when conveying messages related to:

- trends and patterns in data distribution
- large quantities of data
- representation of how values change over time.

There are many charts and graphs available to you, particularly when using a computer, so you need to consider the aims of the chart or graph in order to get the correct message across to the reader.

Line graphs

Line graphs are good for showing change with respect to time and they are commonly used for this representation.

Golden rules:

- Gridlines help the reader.

- Use a sensible scale and use the same scale when comparing graphs.

- Always label the axes.

> ### TIP
>
> If the range of y-axis values is small, trends can be harder to see. Try adjusting the y-axis bounds of the graph to better convey the message. You do not need to start the scale at zero. For example, Figure 5.12(a) shows the same data as Figure 5.12(b), but the increase in profits is more obvious.
>
> **Figure 5.12** Profit over time: (a) more obvious, (b) not so obvious

Ⓐ Worked example

Present the data in Table 5.7 in a chart to clearly show the trend of increasing sales over time.

The data in Table 5.7 can be plotted on a line graph, as shown in Figure 5.13. This clearly shows increasing sales over time.

Figure 5.14 represents the same data as Figure 5.13, but the lower bound used is 2 rather than 0 used in Figure 5.13. The result is that the curve is steeper and the message of increasing sales is more significant.

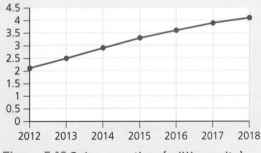

Figure 5.13 Sales over time (million units)

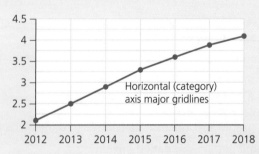

Figure 5.14 Sales over time with a steeper curve

Pie charts

Pie charts are often used badly to represent data. Take the example in Figure 5.15. All the sample values are so close together that on the pie chart it is very difficult to see the differences between 2012, 2013 and 2014, which look similar, and 2015, 2016, 2017 and 2018, which also all look similar. This chart does not convey a message of any real value.

However, a pie chart would be an excellent choice to present the data in Table 5.5. In fact, if the message is to convey the popularity of space grey over any other surface finish, the pie chart achieves this with greater visual emphasis than the original bar graph. In Figure 5.16, the popularity of space grey is resounding and overshadows all the values, carrying greater visual weight and clearly showing the reader that more than half of those sampled chose space grey.

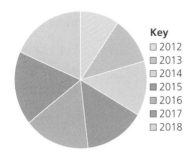

Figure 5.15 Pie chart: sales over time (million units)

B Guided questions

1 **Look back at Table 5.1, which collates the results of a survey of laptops weighing less than 1 kg. Present the findings for weight in an appropriate chart.**

   Step 1: Identify the message you want to convey.

   Step 2: Collate the results into an appropriate form in a table. Use a spreadsheet application to do this. This will enable you to organise and re-order data with ease. For example, most have a function to re-order values into ascending order.

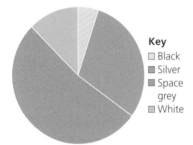

Figure 5.16 Colour preferences: frequency

   Step 3: Use the spreadsheet application chart wizard to create appropriate charts.

2 **Use a chart to identify any trends between laptop screen size and weight.**

   Step 1: Collate the results into an appropriate form in a table. Use a spreadsheet application to do this.

   Step 2: Use the spreadsheet application chart wizard to create an appropriate chart.

5.5 Interpreting and extracting appropriate data

Table 5.5 and Figure 5.16 are not just the visual presentation of data in a form that is easy for the user to read, they also provide numerical data in a simple form for further analysis.

In your examinations, you will be expected to be able to analyse data in forms like these and present quantities as either fractions or percentages of each other or the total.

In your NEA you should analyse your research data and draw conclusions to define the stakeholder requirements. For example, if you were to make only one colour of laptop, you would conclude that 'space grey was the most popular'. However, this statement would have been true if only one more person chose space grey than silver. In this case, it would be tricky to justify one finish over the other and you might need to interview a larger sample or provide both surface finishes in order for the product to be commercially viable. Therefore, it would be better to provide evidence that space grey is significantly more popular than the other finishes.

For example, look back at Table 5.5. You could define how popular space grey was either as a percentage of the total or, more convincingly, in terms of popularity over the next choice, silver. These are included in the worked examples below.

Analysing data using percentages

'Per cent' means per 100, in other words a percentage is a fraction of 100. To convert from a fraction to a percentage you need to multiply by 100 as shown in the worked examples below.

A Worked examples

Look back at Table 5.5 and Figure 5.16.

a **What proportion of the users interviewed preferred the space grey surface finish?**

The simplest way to answer this question is to define the answer as an expression:

$$\frac{21 \text{ users chose space grey}}{40 \text{ users total}}$$

This becomes the fraction $\frac{21}{40}$

This is not a particularly simple fraction, so the answer would be best presented as a percentage. Multiply by 100 to get the percentage.

$$\text{percentage} = \frac{21}{40} \times 100 = 52.5\%$$

Answer: 53% (2 s.f.)

b **How much more popular was the choice of space grey than silver?**

The answer to this question can be defined by either the number of users or as a percentage. The percentage is more useful because it will define space grey relative to silver. Multiply the fraction by 100 to get the percentage.

$$\text{percentage of space grey to silver} \frac{21 \text{ users}}{12 \text{ users}} \times 100 = 175\%$$

Answer: 75% more users chose space grey than silver.

By representing the answer in this way, the significance of the difference between the values is made clear to the reader.

c **What percentage of the users interviewed preferred a surface finish of black?**

This is a percentage of the total.

$$\text{percentage} = \frac{2 \text{ black}}{40 \text{ users total}} \times 100 = 5\%$$

Answer: 5% of the users that were interviewed preferred black.

5.6 Interpret statistical analyses to determine user needs

A design and technology textbook will describe a range of methods by which primary and secondary data can be collected to help determine user needs when designing new products and systems. This chapter describes methods of presenting data through charts and other graphical means, while Chapter 2 deals with how to analyse data in terms of percentages, fractions and ratios.

In this section we will look at more ways in which data can be interpreted. It is worth saying that the same set of data can be analysed and interpreted in different ways and different conclusions can often be drawn. Statistics are, therefore, sometimes used in a somewhat narrow-minded way to argue the case for or against something, as any politician will tell you!

> **TIP**
>
> As a designer, you will need to interpret data to help inform your designs so that they meet user needs and are fit for purpose. It is important to take a balanced view of statistical analyses and understand that it is risky to base a design on firm and unyielding conclusions drawn from statistical data. This is simply because an alternative analysis of the same data may lead to a different conclusion.
>
> One problem that often crops up is where a conclusion is reached based on incomplete statistical data, although this may not be immediately obvious. Also, take care not to search for and collect only data which further support an incorrect conclusion you have already reached.

Data sample

When collecting data to determine user needs, it is usually possible to take data from only a **sample** of potential users rather than from every possible user. For the data to be reliably applied across an extended range of users, it is important that the sampled users are **representative** of the extended set of potential users.

When interpreting secondary data (data collected by someone else) it is usually important to consider how the data were collected or the sample that was used. It is always a risk to draw conclusions based on data from unknown sources without considering whether such data are representative of the identified users in your own design scenario.

Interpreting data

The problems below are examples of the type of data collection and presentation which might be used to determine user needs as part of design and technology NEA projects. The questions which follow should help you to understand the type of analysis that you could carry out to interpret the results so that they provide useful and meaningful information to support your decisions through the designing stages.

Use the techniques explained in Chapter 2 and here to analyse and interpret the data.

A Worked examples

80 people were surveyed and asked to indicate whether they used headphones to listen to music. The results were grouped by age and displayed on the bar chart in Figure 5.17.

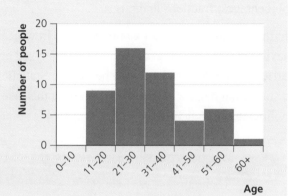

a **Identify the single age range which indicates the largest number of headphone users.**

It is easy to see from the bar chart that more people in the 21–30-year-old group use headphones than any other group.

Figure 5.17 People using headphones

b **Use a calculation to support the conclusion 'Approximately twice as many 21–30 year olds use headphones compared with people aged over 51'.**

16 people use headphones in the 21–30-year-old group.

In the groups aged over 51, (6 + 1) = 7 people use headphones.

The ratio of people in these two groups is $\frac{16}{7} = 2.3$, which indicates that there are approximately twice as many people in the 21–30-year-old group.

c **Calculate the percentage of people in the survey who indicated that they use headphones to listen to music.**

It is necessary to work out the total number of people who use headphones. This is done by adding together the numbers in each group:

Total number of headphone users = 9 + 16 + 12 + 4 + 6 + 1 = 48

The total number of people in the survey is 80. Therefore, the percentage of headphone users is:

$$\frac{48}{80} \times 100 = 60\%$$

d **The largest users of headphones are in the 11–40 age group. Of all the people who use headphones, calculate the percentage that this group occupies.**

The total number of headphone users was calculated in part (c) to be 48.

The number of headphone users in the 11–40 age group is (9 + 16 + 12) = 37.

Therefore, the percentage of users in the 11–40 age group is:

$$\frac{37}{48} \times 100 = 77\% \text{ (rounded to nearest \%)}$$

B Guided question

1 **A design and technology student collects height data from a class of Year 6 students. The data are presented in Figure 5.18.**

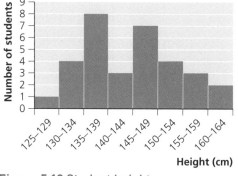

Figure 5.18 Student heights

   a **Calculate the number of students in the class.**

   This is determined by adding together the numbers of students in each height group.

   b **Identify the most common height group in the class.**

   Note that the most common group is also called the **mode**.

   c **Use the bar chart to calculate the height range of students in the class.**

   As the height data in the bar chart are grouped, it is not possible to identify the exact height of the shortest or the tallest student. Therefore, it would be reasonable to calculate the range identified by the two extreme height groups on the bar chart:

   height range = 164 − 125 = 39 cm.

   Note: the extreme values are not recorded because the classes are open-ended, but the values used are reasonable.

   d **Calculate the mean height of students in the class.**

   As the height data are grouped, we can only estimate the mean height of the class by assuming that each student has a height which sits in the centre of each group's range (see Table 5.8).

Table 5.8 Student height ranges

Height range (cm)	125–129	130–134	135–139	140–144	145–149	150–154	155–159	160–164
Centre height (cm)	127	132	137	142	147	152	157	162
Number of students	1	4	8	3	7	4	3	2

   Notice that this value could have been easily *estimated* simply by looking at the shape of the bar graph. In many instances, when determining user needs by interpreting data presented graphically, an estimation is sufficient and is far quicker to determine.

C Practice questions

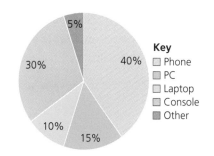

Figure 5.19 Video game devices

2 A designer is investigating the devices people use to play video games. 120 people were surveyed and the results are shown in the pie chart in Figure 5.19.

   a Use the chart to identify the second most used gaming device.
   b Identify the number of people in the survey who use a laptop to play video games.
   c Calculate the ratio of people who use a console to those who use a PC.
   d Calculate the number of people in the survey who would need to change from phone to console to make these two categories equal.

3 An automated voting machine in an airport asked whether visitors were enjoying their day. Those who voted yes were asked to indicate their age. The results were recorded and displayed in Figure 5.20.

a Calculate the total number of people who voted yes.

b Calculate the percentage of yes voters who are in the 30–50 age group.

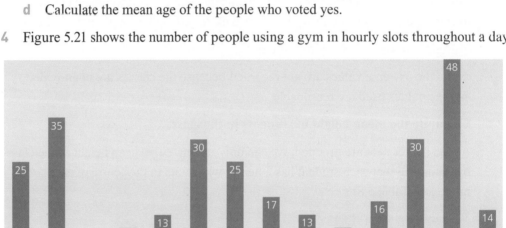

Figure 5.20 People who voted yes, grouped by age

c From Figure 5.20, a design and technology student concludes that the group in the airport who enjoyed their day the most are aged 30–50, while people aged over 90 did not enjoy their day very much at all. Explain an alternative conclusion that could be drawn from this chart.

d Calculate the mean age of the people who voted yes.

4 Figure 5.21 shows the number of people using a gym in hourly slots throughout a day.

Time	Users
7–8	25
8–9	35
9–10	4
10–11	10
11–12	13
12–13	30
13–14	25
14–15	17
15–16	13
16–17	10
17–18	16
18–19	30
19–20	48
20–21	14

Figure 5.21 Gym usage through the day

a Use the chart to identify the busiest hour and the quietest hour during the day.

b Calculate the ratio of users in the busiest and quietest hours.

c Calculate the number of people who use the gym before 9.00 hours.

d Calculate the mean number of users per hour throughout the day.

e Identify the hours that have less than the mean number of users throughout the day.

Sections 5.7 Graphs of motion, 5.8 Engineering graphs and 5.9 Waveforms are available online at www.hoddereducation.co.uk/essentialmathsanswers

6 Coordinates and geometry

6.1 Coordinates in geometric shapes

Two-dimensional coordinates

Coordinates are a set of two numbers which show an exact position on a two-dimensional surface (there are other types of coordinates but, for A-level design and technology, you need only be concerned with the two-dimensional type).

Coordinates are used extensively on graphs, but they are also used to help manufacturers accurately mark out shapes onto flat materials ready for cutting, and they are especially useful when creating shapes using CAD software.

x-axis and *y*-axis

To use coordinates, we need to set up two perpendicular number lines, each called an **axis** (plural '**axes**'). The horizontal axis is the ***x*-axis** and the vertical axis is the ***y*-axis**. The point where the two zeros on the number lines meet is called the **origin**.

Coordinates are written as two numbers contained within brackets and separated by a comma: (x, y), e.g.: (2, 3), (6.0, 4.5), (−4, 6), (7, −2).

These four pairs of coordinates are marked as points on the axes in Figure 6.1. The first number is the distance along the *x*-axis, the second number is the distance up the *y*-axis.

Negative coordinate values indicate that the point lies to the left of the origin on the *x*-axis, or below the origin on the *y*-axis.

> **TIP**
>
> It might help to remember that 'x is a cross', so the *x*-axis is the line that goes *across*.

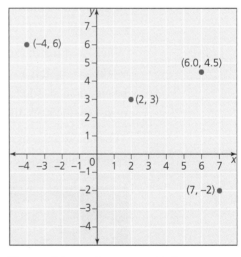

Figure 6.1 *x* and *y* axes, with four coordinate points marked

> **TIP**
>
> To help you remember which coordinate is which, it might help to think that 'you go *along* the corridor before going *up* the stairs', so the first number is the distance *along* the *x*-axis, the second number is the distance *up* the *y*-axis.
>
> $(x, \text{then } y)$ (along, then up) $(\rightarrow, \uparrow)$

Axes and units

When we use coordinates in design and technology, each axis often represents a real quantity, so we need to ensure that we write a unit label after the quantity. For example, if the coordinate represents a distance in mm from an origin, then we might write (30 mm, 45 mm). Alternatively, we could specify that 'all dimensions are in mm', then writing (30, 45) would suffice.

Graphs are often used to represent data in a visual way and, in such cases, the axes will often be representing a physical quantity, so units will be specified on each axis, as in Figure 6.2.

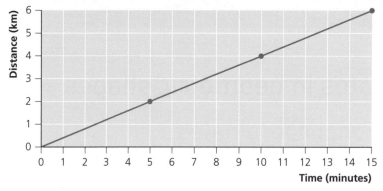

Figure 6.2 A graph with units labelled on each axis

Datum

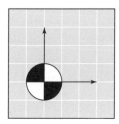

Figure 6.3 Symbol for a datum point

In the same way that the origin of a graph is the point from which all coordinates are measured, a **datum** is a point, or an axis, or a surface on a material from which all measurements are made. The concept is simple – if the datum is accurately known (and never moves), and measurements are always made from this, then all subsequent measurements should be accurate and true.

A datum is sometimes marked as a single point on a material or a drawing, similar to an origin on a graph (see Figure 6.3). x- and y-axes can then be marked out from the datum point using a try square or a T-square tool.

Marking the datum point correctly can be crucial when cutting out patterned materials such as textiles, where correct positioning of the pattern on the part is important.

When marking out onto most materials, it is useful to have two datum edges of the workpiece which are square (perpendicular, at 90°) to each other. If necessary they may be cut, planed, filed or sanded to make them so. In all the marking out that follows, all measurements are then taken from either of the two datum edges.

The photographs show how this works in practice. A rule, a try square and a pencil are used to mark out a rectangle onto a sheet of plywood which does not have four square edges.

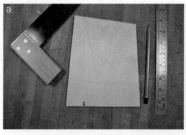  

Figure 6.4 (a) A sheet of plywood has one straight edge that is selected to be the first datum edge, marked with a 1 (b) A try square is used to mark a second datum line at right angles to the first (c) The plywood is planed down to this second datum line to produce a second datum edge (marked with a 2), perpendicular to the first. The width and height of the rectangle are measured from each datum edge and marked with a try square

A CNC machine is a computer-controlled cutting or marking machine, such as a laser cutter or a router. When these machines are first turned on, they reset their cutting head to a known (x, y) datum by moving the head as far as it will travel until it touches a microswitch or a sensor. At this point the position of the head is accurately known and all subsequent movements of the head are then made from this known datum point.

Computer-aided design

Using the rules of geometry that you have learned in GCSE Mathematics, you can quickly plan and draw accurate two-dimensional drawings in computer-aided design (CAD) software. This is a vital skill when using three-dimensional CAD as well, as most three-dimensional shapes begin their life extruded from a two-dimensional profile.

Many designs for design and technology projects consist of regular, geometric shapes (as opposed to free-flowing curves or random-shaped 'blobjects'). Even more complex shapes can often be broken down into a series of connected geometric shapes. In such designs it is recommended that you produce a quick freehand sketch on paper, showing the key dimensions. Be prepared to perform some calculations to work out line lengths (and angles, if needed) as time spent with a pencil and a calculator at this stage can save a great deal of time later. These measurements can then be quickly transferred to your CAD drawing using the principle of x- and y-coordinates.

Producing an accurate drawing on graph paper will make the process of transferring to CAD even easier.

In Figure 6.5, notice the two sets of coordinates, labelled **Abs** (absolute) and **Rel** (relative).

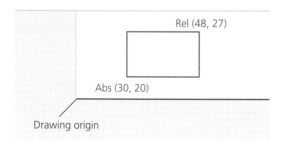

Figure 6.5 2D CAD coordinates can be absolute or relative

Absolute coordinates have their origin at the bottom left of the work area. The rectangle in the figure has its bottom left corner (30, 20) mm from the drawing origin.

Relative coordinates are measured from the last point drawn on the CAD drawing. The top right corner of the rectangle in Figure 6.5 is positioned (48, 27) mm from the bottom left corner. Relative coordinates make it an easy task to draw precise shapes, provided you know the dimensions of the shape.

When drawing shapes, coordinates can be typed into the boxes at the top left corner of the screen:

- Switch the coordinates to relative (Rel).

- Select a drawing tool and click one corner to start drawing the line or shape.

- Type in the x- and y-coordinates to define the opposite end of the line or shape, relative to the start.

- Click OK and the shape will be drawn.

Examples of relative coordinates:

- Rel (50, 0) will draw a horizontal line 50 mm long.

- Rel (0, 100) will draw a vertical line 100 mm long.

- Rel (60, 60) will draw a 45° line (or a square with 60 mm sides, depending on which drawing tool is selected).

The drawing origin on a CAD drawing can be repositioned to a new location and this can be helpful when drawing complex shapes. In such cases, it would be usual to reposition the drawing origin to the bottom left corner of the shape being drawn.

A Worked example

Use coordinates on CAD software to draw the two-dimensional shape shown in Figure 6.6 (all dimensions are in mm).

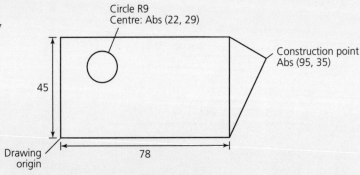

Figure 6.6 Two-dimensional shape

Step 1: Draw the rectangle:

- Using the rectangle tool, click to position the bottom left corner anywhere on the work area.

- Enter the relative coordinates (78, 45) to define the top right corner and to complete the rectangle.

Step 2: Reposition the drawing origin by attaching it to the bottom left corner of the rectangle.

Step 3: Draw the circle:

- Select the circle tool.
- Enter the absolute coordinates (22, 29) to define the centre of the circle.
- Draw the circle with a radius of 9.

Step 4: Draw the triangle:

- Select the construction point tool.
- Enter the absolute coordinates (95, 35) to position the construction point.
- Use the Attach tool to draw two accurate lines from the corners of the rectangle to the construction point.

6.2 Present accurate two-dimensional and three-dimensional drawings

A significant part of being a successful design and technology student involves communicating your ideas to other people, and one of the quickest ways of doing this is through sketches and drawings. Many design and technology students have been heard to say, 'I can't draw', but you do not need to be an artist to produce two-dimensional and three-dimensional design drawings that carry an enormous amount of information. The phrase 'a picture is worth a thousand words' is actually very true.

It is not the purpose of this book to teach you how to develop your drawing skills. However, by applying some mathematical thinking when you produce engineering drawings, you should be able to generate accurate drawn representations of real objects that communicate a lot of useful information.

Three-dimensional drawings are useful for conveying information about the overall form of a design. Two-dimensional drawings are useful for showing layout of parts, cross-sections, or when it is difficult to draw in three dimensions. Three two-dimensional views from front, side and top will carry as much information as a single three-dimensional view, but they require a bit more thought from the viewer to work out how the three views combine into a single three-dimensional object.

Drawing projections

In drawing language, a **projection** is a representation of a three-dimensional object on a two-dimensional surface. A shadow is a familiar type of projection where a three-dimensional object casts its outline shape onto a two-dimensional surface.

There are lots of different types of drawing projections, but we will consider just two types:

- **Orthographic projection**, which is a **multi-view** projection showing an object from (usually) front, side and top views. Simple two-dimensional views of an object are generated. This type is used for **working drawings** (which give instructions for manufacturing).

- **Isometric projection**, which is a single-view projection showing the object at an angle which reveals three sides. A three-dimensional view of the object is created, although the three-dimensional image looks slightly distorted.

In both types of projection, the images do not contain any perspective (so they differ greatly from 'artistic' drawings), which means that measurements can be taken directly from the drawings along any of the axes. This makes them extremely useful when manufacturing products or components.

Orthographic projection

In orthographic projection, an imaginary projector is shone from behind the viewer at right angles to the front face of the object. A view from the front of the object is projected onto a paper surface placed behind the object, parallel to the face being projected. The result is a simple two-dimensional view of that face of the object. This is repeated for the side face and the top face, and three views are then produced, each looking from perpendicular sides. Figure 6.7 shows how this works.

The three views are usually laid out as shown in Figure 6.8 in a style known as **first-angle projection**. Note the drawing symbol that is used to indicate first-angle projection.

The front and side views are sometimes called **elevations** and the top view is sometimes called a **plan** (or bird's-eye) view.

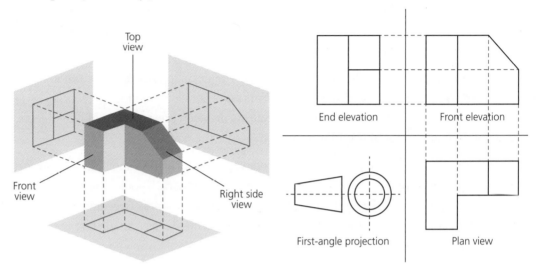

Figure 6.7 Orthographic projection

Figure 6.8 First-angle projection

Two further examples of orthographic first-angle projections of three-dimensional objects are shown in Figure 6.9. Note the dotted lines, which are used to show the 'hidden detail' of the holes. Also note that the top view can be placed under the side view or under the front view, the only difference being the orientation of the top view.

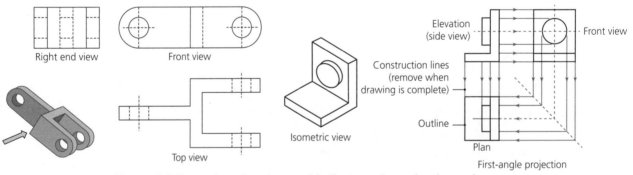

Figure 6.9 Examples of orthographic first-angle projections of three-dimensional objects

---

TIP

When you look at the three orthographic views of an object, try to imagine the object rotating between each view. The rotation takes place along two imaginary hinges on the rear vertical and rear horizontal edges of the object.

Worked example

Draw an orthographic first-angle projection of the shape shown in Figure 6.10.

Consider the three perpendicular faces 1, 2 and 3 in Figure 6.10.

Step 1: Look at the front face (face 1) and project (draw) the two-dimensional view of this face on paper. The shape produced will be the outline shape of the letter 'A'.

Step 2: Take the side face (face 2) and project the view of this face onto the paper, to the left of the front view. The result is a simple rectangle. You cannot tell from the side view alone that the upper part of the side face is sloping away from the viewer.

Step 3: Repeat the projection method for the top face (face 3), placing the resulting top view below the front view (Figure 6.11).

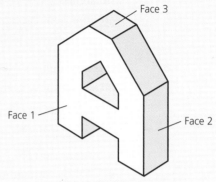

Figure 6.10 Three perpendicular faces

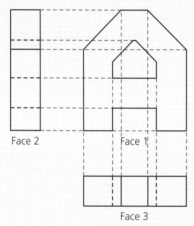

Figure 6.11 Resulting first-angle projection

Guided question

Note: the "B" circle icon precedes "Guided question".

1 **Draw orthographic first-angle projections of the three shapes in Figure 6.12.**

(a)   (b)   (c)

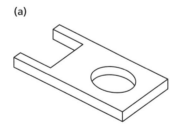 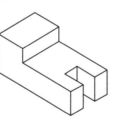 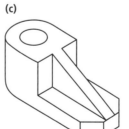

Figure 6.12 Three shapes

You can take measurements directly from the figure and transfer these to your drawings.

Isometric projection

Isometric drawing is an easy method of creating a three-dimensional representation of an object. In isometric drawings, all three-dimensional axes are represented on paper: one vertical axis and two horizontal axes, as in Figure 6.13.

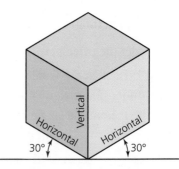

It is a straightforward drawing technique to master providing these mathematical rules are followed:

Figure 6.13 Isometric drawing

- All vertical lines on the object are drawn vertical.
- All horizontal lines on the object are drawn at an angle of 30° to the base line.
- All lines on all the three axes are drawn to the same scale.

Isometric drawings can be produced quickly using **isometric grid paper**, which has lines running vertically and at 30° each side of the horizontal. Alternatively, a 30° set square can be used to help get the angles right when drawing. Experienced designers can produce effective isometric drawings freehand, without the use of grid paper or drawing aids. Practice makes perfect!

Some examples of isometric drawings of objects are shown in Figure 6.14.

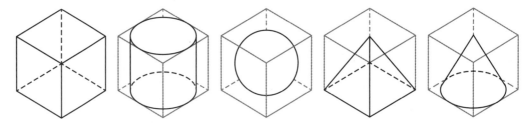

Figure 6.14 Isometric drawings of objects

In isometric drawings, every face is viewed at an angle, which causes the faces to not be their true shape. In the cube, for example, we know that each face is really a square with 90° corners, but if you study the isometric drawing you will notice that this is not the case at all. Nevertheless, our brain interprets the shape as a three-dimensional cube.

This face distortion creates a challenge when drawing circles in isometric because they are not drawn as circles but instead as ellipses. It is frequently necessary to draw circles in design and technology: wheels, cylinders, holes etc. all require circles (or arcs) to be drawn. By far the easiest way to draw isometric circles is to use an **isometric ellipse template**. In use, the ellipse template must be aligned so that the **minor axis** of the ellipse is parallel to the axis of the face on which the circle is being drawn. This takes practice, as it is easy to draw the ellipse in the wrong orientation, which creates an odd-looking isometric circle.

ⓒ Practice question

2   Using isometric grid paper, copy the isometric shapes shown in Figure 6.14.

Orthographic to isometric projection

Visualising or drawing an isometric view from an orthographic first-angle drawing is an important skill to master. It is a straightforward process to convert from two-dimensional orthographic to three-dimensional isometric. Refer to Figure 6.15 and follow the steps below.

- Identify the front, side and top views on the orthographic first-angle drawing.
- Lightly sketch an isometric cuboid on isometric grid paper. Draw the vertical and horizontal dimensions of the cuboid to match the vertical and horizontal dimensions of the orthographic part. Label the front, side and top faces of the cube.
- Transfer the front orthographic view onto the front face of the isometric cube. Remember the three isometric rules:
  - ☐   Vertical lines remain vertical.
  - ☐   Horizontal lines follow the 30° grid lines on the isometric paper.
  - ☐   Line lengths from the orthographic view are transferred to the isometric view with no scaling, i.e. if an orthographic line is 50 mm long then you draw it 50 mm long in isometric (see Tip below).
- Transfer the side orthographic view onto the side face of the isometric cube, following the three isometric rules.
- Transfer the top orthographic view onto the top face of the cube. Notice that the lines on the top orthographic view are horizontal on the object so, in isometric, these will all be drawn at 30°.

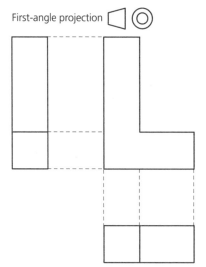

First-angle projection

Figure 6.15 Converting from two-dimensional orthographic to three-dimensional isometric

> **TIP**
>
> Copying orthographic lines full size into isometric produces an isometric drawing that appears approximately 1.22 times larger than the orthographic drawing. If you wish to maintain the same sense of scale, you should reduce the length of every isometric line by a scaling factor of $\frac{1}{1.22} = 0.82$.

Ⓑ Guided question

3   **Sketch an isometric projection of a simple three-dimensional object, e.g. a dice, pen, pencil sharpener, cardboard box etc.**

Step 1: Identify the front, side and top faces of the object.

Step 2: Use a rule to measure the object. Decide whether you intend to draw the isometric projection full size, or with an appropriate scaling factor.

Step 3: Lightly sketch an isometric cuboid which has vertical and horizontal dimensions to match the dimensions of the object.

Step 4: Look at the object from the front view and transfer the two-dimensional front view onto the isometric front view of the cube, remembering the three isometric rules.

Step 5: Transfer the side and top views of the object onto the isometric cube.

C Practice question

4 Use isometric grid paper to draw isometric projections of the orthographic drawings in Figure 6.16.

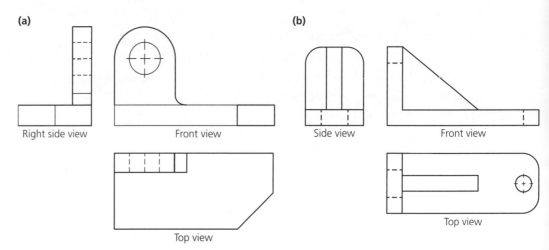

Figure 6.16 **Orthographic drawings**

7 Anthropometrics and probability

7.1 Use data related to human scale and proportion

A lot of design work involves developing products which fit the sizes and shapes of human beings in one way or another. Chairs, desks, bicycles, keyboards, wearable technology and clothing are all examples of products which must 'fit' their human user. Furthermore, as humans come in a variety of shapes and sizes, many products need to be designed to include a large range of human users with varying proportions. Therefore, products may need to be designed to be adjustable, or to be able to accommodate a range of human dimensions without needing modification.

Anthropometric data

Anthropometric data are a large collection of measurements and statistics relating to the human body. Such data are invaluable for designers to use and they reduce the need for a designer to collect such measurements directly.

The word 'anthropometric' comes from the Greek *anthropos* (meaning 'human') and *metron* (meaning 'measure').

Figure 7.1 shows an extract of the anthropometric data relating to a standing human. Notice how the data are grouped:

- Grouped by age range:
  - □ 5–9 years (child), 13–18 years (adolescent), 19–65 years (adult).
  - □ Data would also be available for children under 5 years and adults over 65, should a designer be developing a product for babies, toddlers or the elderly.
- Grouped by gender:
  - □ The adult group is divided into data for men and data for women as there is variation between these two groups.
  - □ For the younger age groups, the male–female data are combined as there is less variation between the sizes of boys and girls, especially in younger children.
- Grouped by percentiles (explained below).

Dimensions (mm)	Age range 5–9 Combined (Percentiles)			Age range 13–18 Combined (Percentiles)			Age range 19–65 Men (Percentiles)			Women (Percentiles)		
	5%	50%	95%	5%	50%	95%	5%	50%	95%	5%	50%	95%
1 Height	1058	1264	1483	1470	1685	1857	1630	1745	1860	1510	1620	1730
2 Eye level	895	1055	1180	1456	1570	1740	1520	1640	1760	1410	1515	1620
3 Shoulder height	843	1014	1198	1184	1352	1525	1340	1445	1550	1240	1330	1420
4 Elbow height	610	720	805	945	1005	1170	1020	1100	1180	950	1020	1090
5 Hip height	496	619	754	734	855	990	850	935	1020	750	820	890
6 Knuckle height (fist grip height)	375	480	565	690	720	815	700	765	830	670	720	770
7 Fingertip height	298	390	470	420	620	695	600	675	730	560	620	680
8 Vertical reach (standing position)	1241	1521	1820	1758	2033	2220	1950	2100	2250	1810	1940	2070
9 Forward grip reach (standing)	442	531	640	594	689	809	720	790	860	660	725	790

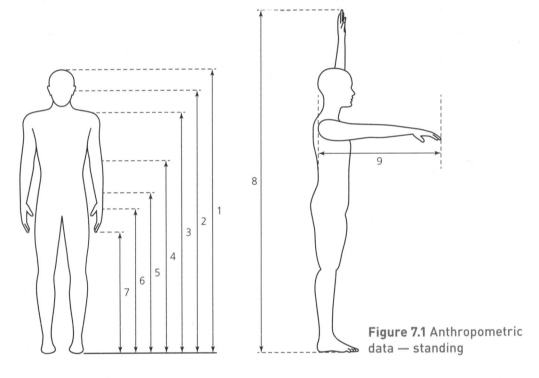

Figure 7.1 Anthropometric data — standing

Percentiles

When height data for a large population of people are collected and presented in a graph, the result in Figure 7.2 is obtained.

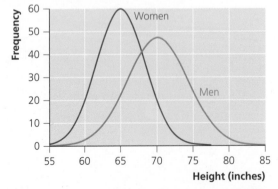

Figure 7.2 Heights of a population naturally form a normal distribution

Full worked solutions at **www.hoddereducation.co.uk/essentialmathsanswers**

This natural spread of heights is called a **normal distribution**. The graph shows that some people are very tall, some people are very short, but the majority of people lie in the centre region of the graph. The most common height of the population coincides with the peak of the curve. Since the curve is symmetrical, the most common height is also the mean height of the population. The same shape of graph is produced for every other body part, such as head circumference, or finger length etc.

Figure 7.3 shows the normal distribution human height curve with 5th, 50th and 95th **percentiles** marked on.

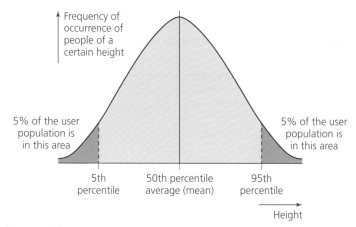

Figure 7.3 Height distribution with 5th, 50th and 95th percentiles marked

- 5% of the population are below the 5th percentile height.
- 5% of the population are above the 95th percentile height.
- 90% of the population are between the 5th percentile height and the 95th percentile height.
- The mean (average) height of the population is the 50th percentile height.

Table 7.1 shows how this applies to anthropometric height data for adult men.

Table 7.1 Height of adult men

5th percentile (1630 mm)	50th percentile (1745 mm)	95th percentile (1860 mm)
5% of the population are shorter than 1630 mm	1745 mm is the most common height in the population 50% of the population are shorter than 1745 mm	95% of the population are shorter than 1860 mm
95% of the population are taller than 1630 mm	50% of the population are taller than 1745 mm	5% of the population are taller than 1860 mm
90% of the population are between 1630 mm and 1860 mm		

The reason designers are interested in the 5th and 95th percentile values is that, by considering these, they are effectively designing their products to be suitable for 90% of the population. This is known as **inclusive designing**. This obviously means that 10% of the population are potentially excluded, but the product will be suitable for most users.

For the 10% of users who fall outside the 5th and 95th percentiles it is sometimes necessary to design particularly small or large versions of the product. Most high street shops, for example, stock shoes only up to UK size 12. Sizes 13 and above often need to be specially ordered. In other cases, adaptors may be needed to make a product suitable for particularly small or large users. Children's car booster seats are an example of this, as they adapt an adult-sized seat to be safe for a child to use.

> **TIP**
>
> 'Percentile' is sometimes shorted to 'centile'. '5th, 50th and 95th' percentiles are sometimes written as '5%, 50% and 95%'. The meaning is exactly the same in both cases.

Worked examples

Anthropometric adult height data (in mm) are shown in Table 7.2.

Table 7.2 Anthropometric adult height data (in mm)

Men			Women		
5th	50th	95th	5th	50th	95th
1630	1745	1860	1510	1620	1730

a **Find the total height range that a designer must consider to include 90% of men and 90% of women.**

90% of the population correspond to the 5th and 95th percentiles. However, the 95th percentile for women lies above the 5th percentile for men, so the two ranges overlap. Therefore, the height range needing consideration is:

- from the 5th centile height for women to the 95th centile height for men
- from 1510 mm to 1860 mm.

b **Calculate the mean (average) height of the adult population if there are equal numbers of men and women.**

The mean height is the 50th percentile value. As there are equal numbers of men and women in the population, the average will be found by adding the 50th percentile values for men and for women and then dividing by 2:

$$\text{average height of population} = \frac{(1745 + 1620)}{2} = 1682.5\,\text{mm}$$

c **Find the height of an aircraft cabin which will allow 95% of men to stand without bumping their heads.**

This question is simply asking for the 95th percentile height for men which, from Table 7.2, can be seen to be 1860 mm. If the cabin is lower than this, then less than 95% of men will be able to stand.

B Guided questions

1 **The height of a door opening is chosen to be equal to the 95th percentile height of a population. In a population of 20 million, calculate the number of people who will need to duck their head to avoid banging it on the doorway.**

If the door opening height is equal to the 95th percentile height, then 5% of people will be taller than this height and will need to duck their heads. Calculate 5% of the 20 million population.

2 **Figure 7.4 shows a prototype handheld controller and anthropometric data for a human hand.**

	Men (percentiles)			Women (percentiles)		
	5%	50%	95%	5%	50%	95%
Thumb length	44.0	51.0	58.0	40.0	47.0	53.0
Thumb width	20.0	23.0	26.0	16.0	19.0	22.0

Figure 7.4 Prototype handheld controller and anthropometric data for a human hand

a **Research indicates that distance *A* in Figure 7.4 should be 0.8 times the user's thumb length. Calculate distance *A* for the average man.**

The 50th percentile value is the average value for the population.

b **The diameter of button B should be no smaller than half the user's thumb width. Calculate the minimum diameter of button B to make it suitable for 95% of women users.**

Since we are looking for a minimum value for the diameter of button B, it is necessary to look at the largest users' thumbs. This is equivalent to the thumb width of the 95th percentile user. Users with smaller thumbs will find that the button is larger than half their thumb width, which is acceptable.

C Practice questions

3 A man's height is 1910 mm. Using the anthropometric height data from the worked example above, explain how you know that more than 95% of the population are shorter than him.

4 A bicycle designer wishes to produce a bike with an adjustable seat height. The ideal seat height is equal to the rider's hip height. The designer would like the seat adjustment range to suit 90% of female users.

a Use the anthropometric data in Figure 7.1 to determine the range over which the seat height needs to be adjustable for 90% of women users.

b Explain why this seat height range would result in the bike being unsuitable for more than half of the potential male users.

7.2 Understand dimensional variations in mass-produced components

Accuracy and precision of measurements

To a manufacturer, or engineer or scientist, nothing can be measured to perfect precision, as every measuring instrument produces a reading with an element of uncertainty. Therefore, there needs to be an acceptance that the measurement produced may be slightly over or under the true value.

When using measuring instruments, the words **accuracy** and **precision** have different meanings:

- An **accurate** instrument will give a reading that is very close to the true reading.
- A **precise** instrument will give lots of decimal places on the reading, but the reading itself may not be very accurate, i.e. it may not be close to the true value.

Accuracy is usually expressed as a 'plus or minus' (±) percentage of the instrument's reading, e.g. ±10%. This means that the true measurement may be up to 10% above or up to 10% below the value specified by the instrument.

The **precision** of an instrument is generally determined by the scale markings on the instrument, or the number of decimal places shown on the digital display. For example, a digital weighing scale may display a reading of 5.38 g with a precision of ±0.01 g.

Perhaps the most basic instrument for measuring length is the **rule**, or tape measure, ideal for larger measurements. Rules used in design and technology generally have markings at 1 mm intervals, giving them a precision of ±0.5 mm. This precision arises because any attempt to judge a measurement between the mm markings would be subjective, i.e. down to the opinion of the user, and the definitive measurement would be precise only to the nearest mm mark which, in the worst case, could be up to 0.5 mm above or below the judged value.

A rule has a precision of ±0.5 mm.

In Figure 7.5 the length of the object is between 23 mm and 24 mm. A user might judge that the measurement is 23.5 mm, so it should be recorded as 23.5 ±0.5 mm.

Figure 7.5 Rule used to measure an object

This precision can be expressed as a percentage of the reading:

$$\frac{0.5}{23.5} \times 100 = 2.1\% \text{ (to 1 d.p.)}$$

So, the reading could be written as 23.5 mm ±2.1%.

Worked example

State the length and percentage uncertainty of the object in Figure 7.6.

Step 1: Write down the length of the object:

Object length = 3.3 mm.

Step 2: Write down the precision of the rule:

Rules with mm markings have a precision of ±0.5 mm.

Step 3: Calculate the percentage precision:

Precision as a % $= \dfrac{0.5}{3.3} \times 100 = 15\%$ (rounded to nearest %).

The object measures 3.3 mm ±15%.

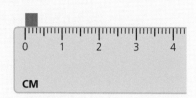

Figure 7.6

> **TIP**
>
> Rules have poor precision for small measurements. To measure small objects (less than, perhaps, 20 mm), consider using one of these high-precision instruments:
>
> - **Vernier caliper** (or a digital caliper) — these versatile devices can usually measure up to 150 mm with a precision of ±0.02 mm.
> - **Micrometer** — these can measure up to about 25 mm to a precision of ±0.005 mm.

Manufacturing tolerance

No manufactured part can be made to perfect precision and, over a large production run, apparently 'identical' parts may differ slightly in dimensions due to wear in machinery, or temperature changes, or operator error etc. The precision (or tolerance) to which a part needs to be made depends on its function in the end product. Mechanical components often need to be manufactured very accurately (known as **close tolerance**), otherwise they may not move efficiently when used in a mechanism; they may create excessive friction if they are too tight, or rattle if they are too loose.

The tolerance of a part specifies the range of sizes over which the part would be acceptable.

- Tolerance may be expressed as a percentage of the measurement, e.g. 5.00 mm ±1%.
- Alternatively, tolerance may be expressed as an actual quantity, e.g. 5.00 mm ±0.05 mm.

In both the above cases, the tolerances are identical, since 1% of 5.00 mm is equal to 0.05 mm.

Tolerance is also considered during quality control where parts may be checked for size and rejected if they are outside an acceptable tolerance.

B Guided questions

1 **A design and technology student uses a 330 Ω resistor with a tolerance of ±5%. Calculate the range within which the actual resistance may lie.**

Step 1: Find 5% of 330 Ω.

Step 2: Use this value to find the upper and lower limits of the resistance.

2　A working drawing specifies that a drive shaft needs to be 8.00 mm with a tolerance of ±0.1 mm. An engineering catalogue contains a shaft of 8.00 mm with a tolerance of ±2.5%. Perform a calculation to decide whether the shaft in the catalogue will definitely be suitable, or not.

Calculate the tolerance on the 8.00 mm catalogue shaft measurement.

Tolerance allowance when manufacturing

In manufacturing industries, tolerance is sometimes used as a 'safety margin', where a fixed value is added to a measurement to make allowance for errors during cutting or to allow for the thickness of the cutting saw blade.

In Figure 7.7, six 80 mm equilateral triangles are tessellated onto a sheet of plywood to make efficient use of the material and to minimise waste. The plywood is to be cut using a bandsaw.

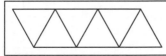

Figure 7.7 Equilateral triangles tessellated onto a sheet of plywood

However, the bandsaw blade has a cutting thickness (a kerf) of about 3 mm. Therefore, each triangular piece would end up slightly smaller than the actual size required. To avoid this, the shapes would be marked out with a small tolerance gap in between each one (probably about a 5 mm gap in this instance), as shown in Figure 7.8. In practice, the bandsaw would probably cut in the gap and then the wooden parts would be sanded down to size.

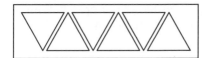

Figure 7.8 Tolerance gap

ⓒ Practice questions

3　Calculate the percentage tolerance of a dimension specified as 400 mm ±5 mm.

4　A student is filing a steel part to a specified length of 80 mm ±2%. Calculate the minimum permitted length of the steel.

5　A 13 mm diameter hole needs drilling to a tolerance of 4%. Calculate the tolerance in mm.

6　A part being turned on a lathe needs to be 18.0 mm diameter with a tolerance of 5%. The part is measured to be 17.2 mm. Calculate whether this is within tolerance.

7　A student is weighing out hardener to mix with polyester resin. The digital weighing scale is accurate to ±2%. Calculate the range of uncertainty in a displayed reading of 10.00 g.

8　Patches like the one shown in Figure 7.9 are to be batch produced from denim fabric. Calculate the minimum dimensions of denim required to tessellate 40 patches in a pattern 5 across and 8 down, allowing for a 10 mm tolerance gap between each patch and at the edges of the material.

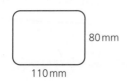

Figure 7.9 Patch to be batch produced from denim fabric

7.3 Probability of defects in batches and reliability

Probability is a way of mathematically describing how *likely* something is to happen, without actually being able to guarantee that it will (or won't) happen. For example, if you toss a coin there is an equal probability that it will land on 'heads' as there is that it will land on 'tails'. However, no one can reliably predict the actual outcome until it happens; in other words, the result is **random**.

Probability has wide-ranging applications relating to topics that a design and technology student may encounter, including statistics, game playing, computer science, artificial intelligence, business finance, manufacturing and physics. For a designer, probability is useful for predicting how product users are likely to behave, or how a computer program will function, or for predicting reliability and failure rate in products.

Defects in batches and reliability of products

Probability theory is useful for helping manufacturers to predict defects in batch production and, consequently, for predicting the reliability of a product. Quality assurance relies on knowing the probability for manufacturing defect at each stage of production and taking steps to compensate for likely defects. In a simple sense, if a manufacturer is able to predict how many products are likely to fail quality control checks, they can arrange to over-manufacture to compensate. Similarly, a purchaser can over-buy parts if they know the probability of the stock being faulty, so that they are not left short of stock when it is needed.

Knowing the probability of faults helps a designer to predict the reliability of a product and also helps them to produce a servicing schedule for engineered products where reliability and failures can be predicted to some extent.

Probability

The mathematical probability of an outcome happening is defined as:

$$\text{probability} = \frac{\text{number of ways the outcome can happen}}{\text{total number of possible outcomes}}$$

Probability is a number between 0 and 1:

- 0 means 'impossibility' (the event will never happen).
- 1 means 'certainty' (the event will definitely happen).
- 0.5 means that there is an even (50 : 50) chance that the event will happen.

Probability can be written as a decimal (0.5), a fraction $\left(\frac{1}{2}\right)$ or a percentage (50%).

Worked examples

a **A dice is rolled to generate a random number between 1 and 6. Calculate the probability of rolling a 5.**

$$P(5) = \frac{\text{number of ways of rolling a 5}}{\text{total number of outcomes}} = \frac{1}{6} \text{ (or 0.167 (to 3 d.p.))}$$

b **85 M4 screws and 37 M5 screws have been mixed up in a box. Calculate the probability of randomly picking out an M5 screw from the box.**

There are 37 chances of picking an M5 screw.

There are (85 + 37) = 122 possible screws to pick.

Therefore, probability of picking an M5 screw is:

$$P(M5) = \frac{\text{number of M5 screws}}{\text{total number of screws}} = \frac{37}{122} = 0.303 \text{ (to 3 d.p.)}$$

c **In a batch of 3000 resistors, the manufacturer expects 15 to be faulty. Calculate the probability of picking one faulty resistor from the batch.**

$$P(\text{faulty}) = \frac{\text{number of faulty resistors}}{\text{total number of resistors}} = \frac{15}{3000} = 0.005$$

d **In a batch of zip fasteners, the probability of a faulty fastener is 0.012. Calculate the likely number of faulty zips in a sample of 500.**

$$P(\text{faulty}) = \frac{\text{number of faulty zips}}{\text{total number of zips}}$$

Therefore,

$$\text{number of faulty zips} = P(\text{faulty}) \times (\text{total number of zips})$$

$$= 0.012 \times 500$$

$$= 6$$

Mutually exclusive outcomes

It is often useful to work out the probability that either one outcome or another outcome will occur. When the two outcomes cannot occur at the same time, they are called **mutually exclusive**. Rolling a dice and hoping for a 3 or a 4 is an example of generating mutually exclusive outcomes because, while six numbers can be generated on each roll, only one number can be generated at a time.

If the two possible outcomes are labelled A and B, then the probability of either outcome happening is:

probability of (A *or* B) = probability of (A) + probability of (B)

When there are only two possible outcomes (such as when tossing a coin):

probability of (A happening) + probability of (A *not* happening) = 1

Ⓑ Guided question

1 **Ten green LEDs, five red LEDs and three blue LEDs are mixed up in a box. If a single LED is randomly picked from the box, calculate the probability that it is green or red.**

Step 1: Calculate the probabilities for individually picking a green LED or a red LED.

Step 2: It is not possible to pick a single green and a red at the same time, so the two outcomes are mutually exclusive.

Non-mutually exclusive outcomes

Non-mutually exclusive means that two outcomes can happen at the same time. Examples of this are:

- tossing two coins and hoping for either to be heads — both could land on heads together
- rolling two dice and hoping for either to be a 3 — both could be a 3 together
- hoping for tomorrow's weather to be dry or warm — it is possible to be dry and warm at the same time.

When events are not mutually exclusive, adding their probabilities causes the outcome where they occur together to be counted twice, so it is necessary to subtract the probability of the outcomes occurring together.

For events that are *not* **mutually exclusive**, the probability that *either* outcome will happen is:

$$P(A \text{ } or \text{ } B) = P(A) + P(B) - P(A \text{ } and \text{ } B)$$

 ## Worked example

A component supplier offers four lengths of screws (10 mm, 20 mm, 30 mm and 40 mm), each in a choice of three materials (steel, stainless steel and brass). Given that each screw has an equal probability of being selected, calculate the probability that a customer will randomly choose a 40 mm screw or a brass screw.

There are four different screw lengths, so the probability of selecting 40 mm is:

$$P(40 \text{ mm}) = \frac{1}{4}$$

There are three different materials, so the probability of selecting brass is:

$$P(\text{brass}) = \frac{1}{3}$$

These two possible outcomes are not mutually exclusive, as it is possible to select a screw which is both 40 mm long and made from brass. Therefore, use the probability formula:

$$P(A \text{ } or \text{ } B) = P(A) + P(B) - P(A \text{ } and \text{ } B)$$

The probability of selecting a screw which is 40 mm *and* brass is $\frac{1}{12}$ as there are $(4 \times 3) = 12$ possible screw types. Therefore:

$$P(40 \text{ mm } or \text{ brass}) = \frac{1}{4} + \frac{1}{3} - \frac{1}{12}$$

$$= \frac{1}{2}$$

There is a 50% probability that a customer will select a 40 mm screw or a brass screw.

Independent events

Two random events are **independent** when the first event does not influence the probability of the second event. Examples of independent events are:

- landing on heads after tossing a coin and then rolling a 6 on a dice
- randomly choosing a paint colour and then randomly choosing a paintbrush size.

When two events are independent, the probability of *both* happening together is:

$$P(A \text{ } and \text{ } B) = P(A) \times P(B)$$

Dependent events

Two events are **dependent** if the outcome of the first event affects the probability of the second event. This is more common than you might expect. Examples of dependent events are:

- picking a 'king' from a pack of cards and then picking an 'ace' from the same pack. The probability of selecting the first king is 4/52, but if it is not replaced in the pack, the probability of picking an ace is now 4/51
- choosing one of 5 working parts from a batch of 10, then choosing another working part from the same batch. The probability of choosing the first working part is 5/10 (= 0.5), but once this part is removed, the probability of choosing another working part has reduced to 4/9 (= 0.444).

For two dependent events, the probability of *both* happening together is:

$$P(A \text{ and } B) = P(A) \times P(B \text{ given } A)$$

The term $P(B \text{ given } A)$ means the probability of event B happening, given that event A has already happened.

For the example above, the probability of selecting two working parts from the batch of 10 would be, therefore:

$$P(A \text{ and } B) = \frac{5}{10} \times \frac{4}{9} = \frac{20}{90} = \frac{2}{9}$$

Ⓐ Worked examples

a Two coins are tossed.

i Calculate the probability that *both* land on heads.

The two coin tosses are independent events.

The probability of each coin landing on heads is $\frac{1}{2}$.

Therefore, the probability of one coin *and* the other landing on heads is:

$$P(A \text{ heads } and \text{ B heads}) = \frac{1}{2} \times \frac{1}{2} = \frac{1}{4} \text{ (or 0.25)}$$

ii Calculate the probability of *either* coin landing on heads.

The events are not mutually exclusive, as both coins could land on heads together. Therefore, the probability of either coin landing on heads is:

$$P(A \text{ heads } or \text{ B heads}) = P(A \text{ heads}) + P(B \text{ heads}) - P(A \text{ heads } and \text{ B heads})$$

$$= \frac{1}{2} + \frac{1}{2} - \frac{1}{4} = 0.75$$

b Quality control checks on a batch of 200 shirts suggest that one shirt in 25 is faulty.

i Calculate the probability of finding a faulty shirt in the batch.

$$P(\text{faulty}) = \frac{1}{25} = 0.04$$

ii Two shirts are selected at random from the batch. Calculate the probability that both are faulty.

The selections of the two shirts are dependent events. In the batch of 200, there is a likelihood of $(200 \times 0.04) =$ eight faulty shirts.

The probability of selecting the first faulty shirt is:

$P(A) = 0.04$

Given that the first shirt is faulty, the probability of then selecting another faulty shirt is:

$P(B \text{ given } A) = \dfrac{7}{199} = 0.035 \text{ (to 3 d.p.)}$

Therefore, the probability of selecting two faulty shirts is:

$$P(A \text{ and } B) = P(A) \times P(B \text{ given } A)$$
$$= 0.04 \times 0.035$$
$$= 0.0014$$

C Practice questions

2   A student is developing a target game. The target has a total area of $1000\,\text{cm}^2$. A bonus section within the target has an area of $20\,\text{cm}^2$. Calculate the probability of hitting the bonus section if the user aims randomly at the target.

3   Twelve blue LEDs, six green LEDs and four red LEDs are mixed up in a box.

   a   Calculate the probability of picking a single green LED from the box.

   b   Calculate the probability of picking either a single green or a single red LED from the box.

   c   Calculate the probability of picking two blue LEDs from the box.

4   A student is developing a simple game consisting of two displays which each independently generate a random number from 1 to 5.

   a   Calculate the probability of each display generating a number 3.

   b   Calculate the probability of each display generating the same number.

5   In a survey of 18 girls and 12 boys, 22 children said that they ride a bike. If one child is selected randomly from the group:

   a   Calculate the probability that the selected child rides a bike.

   b   Calculate the probability that the selected child is a girl who rides a bike.

6   A supplier sells five sizes of batteries: AAA, AA, C, D and PP3. Each size is available in 'normal life' or 'extended life'. If buyers choose batteries randomly, calculate the probability of selling an extended-life AA battery.

7   A component in an electronic system has a 0.1 probability of overheating. If two such components are independently used in the system, calculate the probability that either one will overheat.

8   In a batch of 150 lamps, 25 are known to be faulty. Calculate the probability of randomly selecting three faulty lamps from this batch.

Appendix

Specification coverage

The table below shows the specification for maths content for AS and A-level design and technology in the required maths skill column. These skills are common to all boards. The table also illustrates where these maths skills could be developed or assessed during the course. Skills shown in bold type will only be tested in the full A-level course, not the AS course. It is important that you realise that this list of examples is not exhaustive and these skills could be developed in other areas of specification content from those indicated. The information in the table is intended as a guide only. You should refer to your specification for full details of the topics you need to know.

DE = design engineering; FT = fashion and textiles; PD = product design; ED = engineering design

Topic in this book	Required maths skill	AQA	CCEA	Edexcel	OCR	WJEC
			Option A (1.11–1.17) Electronic and microelectronic control systems, Option B (1.18–1.23) Mechanical and pneumatic control systems, Option C (1.24–1.31) Product design			ED (2.3.1, 2.3.2) -, FT (2.3.3,2.3.4), PD (2.3.5,2.3.6)
1 Using numbers and percentages						
1.1 Units, powers, standard form and accuracy	1.1 Calculation of quantities of materials, components, costs and size with consideration of percentage profits and tolerances as appropriate	3.1.1, 3.1.6.2, 3.2.7	1.1, 1.2, 1.3, 1.4 , 1.5, 1.7	1, 2, 9.1a, 11.2a	DE 1.2a, 5.3b, 8.1c FT 1.2a, 5.3b, 7.2a, 7.2e, 8.1c, PD 2.1a, 5.3b, 7.2e, 8.1c	2.2c, 2.3.3b, 2.3.5a, 2.3.6a
	1.3 Confident use of decimal and standard form	3.1.1	1.1, 1.2, 1.3, 1.4 , 1.5, 1.7	1, 2, 9.1a, 11.2a	DE 6.4a, 6.4b	2.2c, 2.3e, 2.3.1f
1.2 Working with formulae and equations	1.2 Substitute numerical values into and rearrange learnt formulae and expressions	3.1.7	1.14, 1.15, 1.20, 1.22	1, 2, 9.1a, 11.2a	DE 6.4b, 6.4e, 6.2b	2.2c, 2.3e, 2.3.1f

1.3 Equations of motion (OCR Design Engineering only)	1.4 Recall and application of engineering formulae in qualitative work and calculations when applying engineering to mathematical skills		1.14, 1.15, 1.20, 1.22		DE 6.4a, 6.4b, 6.4e, 6.2b	2.3e, 2.3.1f
1.4 Scientific and engineering formulae	1.5 Recall and application of scientific formulae		1.14, 1.15, 1.20, 1.22		DE 6.4a, 6.4b, 6.4e, 6.2b	2.3e, 2.3.1f
2 Ratios and percentages						
2.1 Ratios and scaling of lengths, area and volume	2.1 Understand and use ratios in the scaling of drawings and pattern grading	3.1.14	1.8	3.3	DE 4.1b, FT 7.2a	2.1c, 2.3.4b, 2.3.6a
2.2 Ratios and mechanisms			1.20, 1.22		DE 6.4a, 6.4b, 6.4e, 6.2b	2.3e, 2.3.1f
2.3 Working with percentages	2.4 Calculate percentages e.g. with profit, waste saving calculations or comparing measurements	3.1.13	1.7	3.3, 9.1d	DE 3.1a	2.1b, 2.2c
3 Surface area and volume						
3.1 Properties and areas of two-dimensional shapes	3.1 Determining quantities of materials by surface area	3.1.2, 3.2.7	1.8, 1.26	1, 2, 3.3c, 3.3d, 9.1a	FT 8.1c	2.1c, 2.2c, 2.3.6a
3.2 Surface area of three-dimensional objects	3.2 Calculate the overall surface area of different shapes, such as cuboids, cylinders and spheres to determine quantities of material and feasibility analysis	3.2.7	1.8, 1.26	1, 2, 3.3c, 3.3d, 9.1a	FT 8.1c	2.1c, 2.2c, 2.3.6a
3.3 Volume of three-dimensional objects	3.3 Calculate the volume of different shapes, such as cuboids, cylinders and spheres to determine suitability of objects and products	3.1.7	1.8, 1.26	1, 2, 3.3c, 3.3d, 9.1a	FT 8.1c	2.1c, 2.2c, 2.3.6a

3.4 Density and mass of three-dimensional objects		3.1.7	1.8, 1.26	1, 2, 3.3c, 3.3d, 9.1a	FT 8.1c	2.1c, 2.2c, 2.3.6a
4 Trigonometry						
4.1 Pythagoras	4.1 Calculate the sides and angles of objects to determine structural integrity, marking out and direction of movement	3.2.7, 3.1.4.6	1.8, 1.26	3.3	DE 6.1a, 6.2b, 6.3a PD 6.1b	2.1c, 2.3.1f, 2.3.4b, 2.3.6a
4.2 Sine, cosine and tangent	4.1 Calculate the sides and angles of objects to determine structural integrity, marking out and direction of movement	3.2.7, 3.1.4.6	1.8, 1.26	3.3	DE 6.1a, 6.2b, 6.3a PD 6.1b	2.1c, 2.3.1f, 2.3.4b, 2.3.6a
4.3 Sine and cosine rules	4.1 Calculate the sides and angles of objects to determine structural integrity, marking out and direction of movement	3.2.7, 3.1.4.6	1.8, 1.26	3.3	DE 6.1a, 6.2b, 6.3a PD 6.1b	2.1c, 2.3.1f, 2.3.4b, 2.3.6
4.4 Direction of movement	4.4 Determining projectile motion and direction of movement	3.1.7	1.20		DE 6.3a, 6.3b	2.3.1f
4.5 Resolving force vectors	4.5 Determining how to resolve force vectors		1.20		DE 6.3a, 6.3b	2.3.1f
5 Construction, use and analysis of charts and graphs						
5.1 Presenting data	5.1 Representation of data used to inform design decisions and evaluation of outcomes	3.1.14, 3.2.1	1.7	5.1, 5.2, 5.3, 11.1a	DE 2.1a , PD 1.2b, 3.2a, 3.2b	2.1b
5.2 Statistics		3.1.14, 3.2.1	1.7	5.1, 5.2, 5.3, 11.1a	DE 2.1a , PD 1.2b, 3.2a, 3.2b	2.1b
5.3 Group data, estimates, modal class and histograms		3.1.1, 3.1.13, 3.1.14	1.7	3.3, 5.1, 5.2, 5.3, 11.1a	DE 1.2a, 3.2b, PD 1.3a, 3.2a, 3.2b	2.1b, 2.1c

5.4 Presenting market and user data	5.2 Presentation of market data, user preferences, outcomes of market research as part of product design, fashion and textiles	3.1.14	1.7		5.1, 5.2, 5.3, 11.1a	DE 8.1a, 8.1b, PD 1.2b, 3.2a, 3.2b	2.1b
5.5 Interpreting and extracting appropriate data	5.3 Interpret and extract appropriate data. 2.3 understand and apply fractions and percentages when analysing data, survey responses and user questionnaires given in tables and charts	3.1.1, 3.1.13, 3.1.14	1.7		3.3, 5.1, 5.2, 5.3, 11.1a	DE 1.2a, 3.2b, PD 1.3a, 3.2a, 3.2b	2.1b, 2.1c
5.6 Interpret statistical analyses to determine user needs	7.1 Interpret statistical analyses to determine user needs and preferences	3.1.12	1.7		5.1, 5.2, 5.3, 11.1a	DE 1.2a, 3.3a, 3.3b, PD 3.2a, 3.2b	2.1b
5.7 Graphs of motion	5.4 Present and interpret velocity/ time graphs		1.20			DE 6.4c	2.3.1f
5.8 Engineering graphs	5.4 Present and interpret stress–strain and resistance–temperature graphs		1.20			DE 6.4c	2.3.1f
5.9 Waveforms	5.5 Representation of frequency, period, amplitude and phase		1.20			DE 6.4a, 6.4b	2.3.1g
6 Coordinates and geometry							
6.1 Coordinates in geometric shapes	6.1 Use of datum points and geometry when setting out design drawings, when setting out patterns and within engineering drawings	3.1.14, 3.2.7	1.8, 1.26		3.3a, 3.3b	DE 4.1b FT 6.2b, 7.3a PD 4.1b	2.1c, 2.1d, 2.3.4b, 2.3.6a
6.2 Present accurate 2D and 3D drawings	6.2 Present accurate 2D and 3D graphics to communicate design solutions	3.2.7	1.8, 1.26		3.3a, 3.3b	DE 4.1b, PD 4.1b	2.1c, 2.3.4b, 2.3.6a

7 Anthropometrics and probability

7.1 Use data related to human scale and proportion	7.2 Use data related to human scale and proportion to determine product scale and dimensions and sizes and dimensions of fashion products	3.1.12, 3.2.1	1.7	5.2, 5.3	DE 1.3a, 1.3b, PD 1.3b, 2.1a	2.1c
7.2 Understand dimensional variations in mass produced components	7.2 Understanding of dimensional variations in mass produced components	3.1.7	1.1, 1.10	8.2a	DE 7.2c FT 7.2d, PD 7.2d, 7.3b	2.1b
7.3 Probability of defects in batches and reliability	7.3 Defects in batches and reliability linked to probabilities	3.1.7	1.1	8.2a	DE 7.2c	2.1b

Full worked solutions at **www.hoddereducation.co.uk/essentialmathsanswers**